Physical Chemistry in Action

Physical Chemistry in Action presents volumes which outline essential physicochemical principles and techniques needed for areas of interdisciplinary research. The scope and coverage includes all areas of research permeated by physical chemistry: organic and inorganic chemistry; biophysics, biochemistry and the life sciences; the pharmaceutical sciences; crystallography; materials sciences; and many more. This series is aimed at students, researchers and academics who require a fundamental knowledge of physical chemistry for working in their particular research field. The series publishes edited volumes, authored monographs and textbooks, and encourages contributions from field experts working in all of the various disciplines.

Arieh Ben-Naim

Solvent-Induced Interactions and Forces in Protein Folding

Hydrophobic-Hydrophilic Phenomena

Springer

Arieh Ben-Naim
Department of Physical Chemistry
The Hebrew University of Jerusalem
Jerusalem, Israel

ISSN 2197-4349 ISSN 2197-4357 (electronic)
Physical Chemistry in Action
ISBN 978-3-031-31870-2 ISBN 978-3-031-31871-9 (eBook)
https://doi.org/10.1007/978-3-031-31871-9

This Springer imprint is published by the registered company Springer Nature Switzerland AG
The registered company address is: Gewerbestrasse 11, 6330 Cham, Switzerland

This book is dedicated to my friend and colleague,
Prof. Kesharsingh J. Patil.

Preface

In 1980, I published a book titled *Hydrophobic Interactions*, wherein I summarized all that was known at that time about the hydrophobic ($H\phi O$) effects: solvation and hydrophobic interactions. The more general term "solvent-induced interactions" was studied mainly in connection with the effect of a solvent on the interaction between two charged particles.

Toward the end of the 50s, Kauzmann (1959) introduced the idea of "hydrophobic bonds"—later referred to as "hydrophobic interactions"—to explain the stability of the 3D structure of proteins. This idea dominated the field of protein chemistry in its entirety, including the protein folding problem and the self-assembly of proteins. My contributions to the field were: First, a clear-cut *definition* of the concepts of "$H\phi O$ solvation" and "$H\phi O$ interaction," and second, a new and useful *measure* of both the $H\phi O$ solvation and $H\phi O$ interactions, Ben-Naim (1971a, b, 1972a, b, 1974).

The application of the new definition and new measure of the $H\phi O$ effects led to the establishment of the fact that there is a tendency of the two $H\phi O$ molecules to adhere to each other, and that this tendency is far stronger in aqueous solutions than in any other solvent for which the relevant data was available.

In my book's preface, Ben-Naim (1980), I wrote:

> *...in spite of my research in this field over almost ten years, I cannot confirm that there is at present either theoretical or experimental evidence that unequivocally demonstrates the relative importance of hydrophobic interactions over other types of interaction in aqueous solutions.*

In this quoted paragraph, I did not specify the processes in which the $H\phi O$ interactions were involved, but it was clear from the context of the preface as a whole that I was referring specifically to the processes of protein folding and protein–protein association.

In the 1980s, it was believed that direct hydrogen bonds (HBs) did not contribute significantly to the stability of proteins. This led to the widespread acceptance of the idea that $H\phi O$ interactions are the *most important* factors in determining the stability of proteins.

However, the situation changed dramatically in the late 1980s, upon the discovery of some new hydrophilic ($H\phi I$) effects. It was found that some $H\phi I$ interactions are far stronger than the corresponding $H\phi O$ interactions. Contrary to common belief, it was also ascertained that direct HBs contribute significantly to the stability of proteins; and that the role of $H\phi O$ effects in protein was grossly exaggerated.

Unfortunately, there are still numerous scientists who still subscribe to the old and obsolete ideas regarding the role of HBs and $H\phi O$ effects in protein stability in spite of the overwhelming evidence that there are many more $H\phi O$ interactions, and that each of these is far stronger than the corresponding $H\phi O$ effect.

This book is in some sense a sequel to my previous book which was published in 1980. However, its main message is that for quite a long-time people were misled to believe in the idea of the "most important" $H\phi O$ effects. In a way, this book presents a paradigm shift from $H\phi O$ to $H\phi I$. It will be shown that various $H\phi I$ effects not only explain the stability of the structure of proteins and the speed of the process of protein folding but also effectively expunge the mystery associated with some biochemical processes such as protein folding and self-assembly of proteins.

The first three chapters deal with some basic definitions of solvent-induced effects and narrate the story of the evolution of the ideas about the factors that are important for protein stability; from early ideas about HBs to $H\phi O$ effects and finally to the $H\phi I$ effects. In Chap. 4, we summarized what is known about $H\phi O$ solvation and interactions. It is shown that these $H\phi O$ effects are very different in strength when they are applied to *entire* molecules, and on *groups* attached to a protein. Chapter 5 is dedicated to $H\phi I$ effects. Here we present the evidence regarding their strength based on theoretical, experimental, and simulated calculations.

The last chapter summarizes the implementation of the $H\phi I$ effects in the explanations of the long-lasting problems such as protein folding and self-assembly of proteins.

Jerusalem, Israel

Arieh Ben-Naim
https://ariehbennaim.com

References

1. Kauzmann W (1959) Some factors in the interpretation of protein denaturation. Adv Protein Chem 14:1–63
2. Ben-Naim A (1971a) Statistical mechanical study of hydrophobic interaction, Part I: interaction between two identical non-polar solute particles. J Chem Phys 54:1387–1404
3. Ben-Naim A (1971b) Statistical mechanical study of hydrophobic interaction, Part II: interaction among a set of M identical spherical and non-polar solute particles. J Chem Phys 54:3696–3711
4. Ben-Naim A (1972a) Statistical mechanical study of hydrophobic interaction, Part III: generalization and further applications. J Chem Phys 57:5257–5265
5. Ben-Naim A (1972b) Statistical mechanical study of hydrophobic interaction, Part IV: the behavior of the function Y(R) at short distances. J Chem Phys 57:5266–5269

6. Ben-Naim A (1974) Water and aqueous solutions, introduction to a molecular theory. Plenum Press, New York
7. Ben-Naim A (1980) Hydrophobic interactions. Plenum Press, New York

Acknowledgments

Thanks to Dr. Stewart Durell and to Prof. Kesharsingh Patil and Prof. Jose Angel Sordo Gonzalo who read the manuscript and sent me important comments. Thanks also to Dr. Sandeep Patil for providing the photo of his father K. J. Patil, which I used in the dedication page.

As always, I am very grateful for the graceful help from my wife, Ruby, and for her unwavering involvement in every stage of the writing, typing, editing, re-editing, and polishing of the book.

Acknowledgments

Thanks to Dr. Stewart Durell and to Prof. Kesharsingh Patil and Prof. Jose Angel Sordo Gonzalo who read the manuscript and sent me important comments. Thanks also to Dr. Sandeep Patil for providing the photo of his father K. J. Patil, which I used in the dedication page.

As always, I am very grateful for the graceful help from my wife, Ruby, and for her unwavering involvement in every stage of the writing, typing, editing, re-editing, and polishing of the book.

Contents

Abbreviations

1D	One Dimensional
CD	Cold Denaturation
GPF	Grand Partition Function
HB	Hydrogen Bond
HBed	Hydrogen Bonded
HBing	Hydrogen Bonding
HBs	Hydrogen Bonds
$H\phi I$	Hydrophilic
$H\phi O$	Hydrophobic
Ig	Ideal Gas
PF	Partition Function
PFP	Protein Folding Problem
PMF	Potential of Mean Force
SIF	Solvent-Induced Forces
SII	Solvent-Induced Interactions
SMI	Shannon's Measure of Information
SPT	Scaled Particle Theory

Chapter 1
Introduction, Definitions and Motivations for Studying Solvent-Induced Interactions

In the first part of this chapter we start by defining the *interaction energy* between two or more particles in different media and express these interactions in terms of thermodynamic quantities. In the second part, we discuss the motivation for studying these interactions in the context of the so-called protein folding problem (PFP). We shall survey the main historical landmarks in the history of studying solvent-induced interactions (SII) in connection with the PFP.

1.1 Definitions; Direct and Indirect Interactions

The basic idea of solvent-induced interactions (SII) is quite simple. We can view the SII as an extension of the concept of intermolecular interaction between two atoms or two molecules in a vacuum.

Figure 1.1 shows two spherical particles (A and B which can either be identical or different particles). The intermolecular interaction between these two particles is defined as the work (i.e. the force, times distance) involved in bringing the two particles from infinite separation to the final distance R. We denote this work by:

$$W^{(v)}(\infty \to R) = U(R) - U(\infty) = U(R) \tag{1.1}$$

The superscript (v) is for "vacuum," or in an ideal-gas phase. On the left-hand side, we write the work (W) associated with the process, denote by $(\infty \to R)$, as the difference between the potential energy of this system at R, minus the potential energy of the same system at infinite separation (∞). Since the latter is assumed to be zero, we omit $U(\infty)$ and simply identify the work $W^{(v)}(\infty \to R)$ with the potential energy $U(R)$ which is also referred to as the *direct* interaction energy between the two particles at a distance R in a vacuum.

A. Ben-Naim, *Solvent-Induced Interactions and Forces in Protein Folding*, Physical Chemistry in Action, https://doi.org/10.1007/978-3-031-31871-9_1

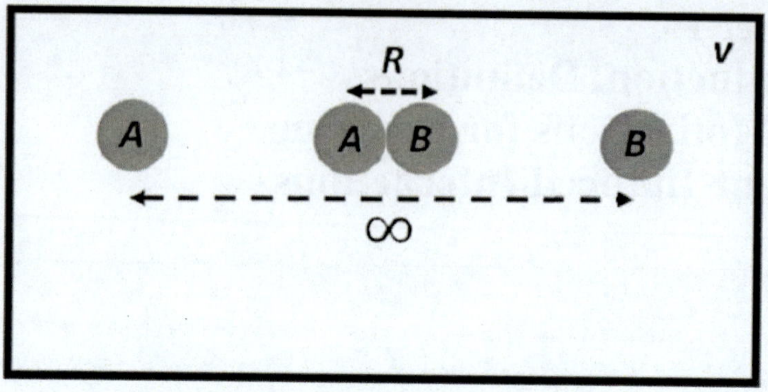

Fig. 1.1 Two simple spherical particles A and B are brought from infinite separation to the final distance R, in a vacuum or in an *ideal-gas* phase

For non-spherical particles, the interaction energy might also be dependent on the orientations of the two particles as well as on the internal configuration of the two molecules, (say, proteins).

We now extend the concept of interaction between the two particles and apply it to the same two particles but in a solvent, or in a liquid phase (l), Fig. 1.2. The work associated with the process ($\infty \to R$) no longer depends only on R but also on some thermodynamic parameters of the system.

Figure 1.2 shows the same process as in Fig. 1.1 but now the two particles are immersed in some "solvent." Clearly, the work associated with the process would depend on the density of the solvent molecules, the temperature at which the process is carried out, and on whether the solvent is or is not at equilibrium.

In this book, we always assume that the solvent is at equilibrium. Also, we assume that process ($\infty \to R$) is carried out very slowly (quasi-statically) in such a way that while the two particles are moved, the state of equilibrium of the solvent is maintained.

The origin of the difference in the amount of work associated with the processes in Figs. 1.1 and 1.2 is easy to understand, qualitatively. Let us start with two simple

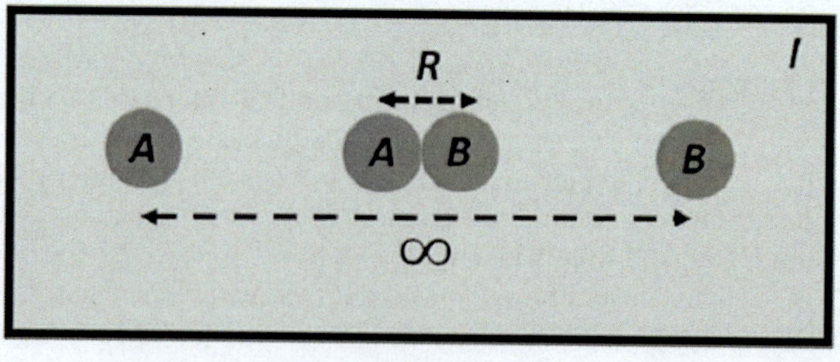

Fig. 1.2 The same process as in Fig. 1.1, but the two spherical particles A and B are in a liquid phase

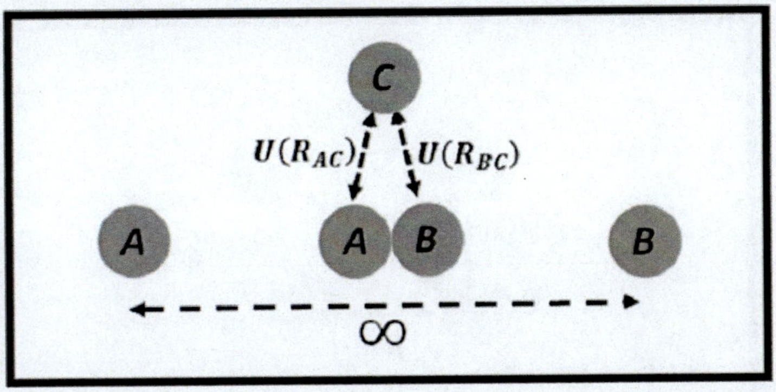

Fig. 1.3 The same process as in Fig. 1.1, but the "solvent" consists of a single particle C

cases of "solvents." Suppose for simplicity that there is only one solvent particle in the system, Fig. 1.3.

In this case, in addition to the *direct* interaction $U(R)$ between the two particles, A and B (which is presumed to be the same as in Fig. 1.1), we also have *indirect* interactions mediated through the "solvent." In the simplest case of Fig. 1.3, we have only one solvent molecule, denoted by C. Here, we have in addition to the interaction $U(R_{AB})$, also the two additional interactions $U(R_{AC})$ and $U(R_{BC})$. These two interactions can either cause attraction or repulsion between the two particles A and B, depending on the type of interactions $U(R_{AC})$ and $U(R_{BC})$, and on the location of the solvent particle. For instance, if A attracts C, and C attracts B, then these two interactions cause an *indirect attraction* between A and B. It should be noted that in a liquid of hard-sphere particles, all the *direct* pairwise interactions are repulsive, yet these repulsive interactions produce *indirect attractive* interactions. For details see Refs. [1–3].

Remember that the process we carried out is the same as in Fig. 1.1. The "solvent" particle is free to move around while the two particles A and B approach each other. Therefore, in studying the work $W(\infty \to R)$ we need to account for the *average* indirect interactions over all possible configurations of the solvent molecule.

Another simple example is shown in Fig. 1.4. Here we have a "solvent" consisting of two, denoted by C and D. In this case, we have in addition to the *direct* interaction $U(R_{AB})$, five additional interactions $U(R_{AC})$, $U(R_{AD})$, $U(R_{BC})$, $U(R_{BD})$ *and* $U(R_{CD})$. These five interactions can either cause attraction or repulsion between the two particles A and B. Again, recall that the process we carried out is the same as in Fig. 1.1. The "solvent' particles are free to move around while the two particles A and B approach each other. Therefore, in studying the work $W(\infty \to R)$ we need to account for the *average* indirect interactions over all possible configurations of the solvent molecules.

Clearly, when the solvent consists of many molecules, the difference between the *direct* interaction $U(R)$, and the *work function* $W(\infty \to R)$, becomes more and more complicated. This is *a fortiori* true when the solvent is a complicated liquid like water. It is here that statistical mechanics is most useful in studying the SII between two (or more) particles (or groups of proteins) in different solvents.

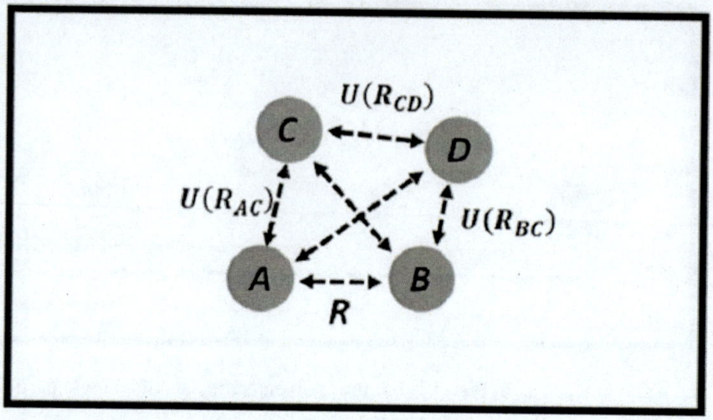

Fig. 1.4 Two particles A and B at fixed positions and at a distance R, in a "solvent" consists of two particles C and D. Here, in addition to the *direct* interaction between A and B, there are five more interactions that contribute to the *indirect* interaction between A and B

In most of this book, we will assume that the solvent is maintained at equilibrium and at a constant volume and temperature. Also, the total number of solvent molecules will always be assumed to be constant. In such a case, the work $W(\infty \to R)$ is related to the change in the Helmholtz energy change in a system consisting of two particles at fixed positions and at a distance R, and a solvent of N molecules in a volume V, and at temperature T. Thus, we write:

$$W^{(l)}(\infty \to R) = A(T, V, N; \infty) - A(T, V, N; R) \tag{1.2}$$

The superscript (l) is for the "liquid phase." In most of the experiments, in which SII has been studied the solvent was maintained at constant temperature and pressure (rather than volume). In this case, the work $W(\infty \to R)$ is related to the Gibbs energy change in the process, i.e.

$$W^{(l)}(\infty \to R) = G(T, P, N; \infty) - G(T, P, N; R) \tag{1.3}$$

Fortunately, for the processes we will be discussing in this book, there is not much difference in the numerical values of the Helmholtz and Gibbs energies, see Ben-Naim [2], for some numerical examples.

Figure 1.5 shows the same process $(\infty \to R)$, as discussed above carried out in both an ideal gas (or in a vacuum) and in a liquid phase (l). We now define the solvent-induced interactions (SII) by difference:

$$\begin{aligned} \delta W(R) &= W^{(l)}(\infty \to R) - W^{(v)}(\infty \to R) \\ &= \Delta A(R) - U(R) \\ &= \Delta G(R) - U(R) \end{aligned} \tag{1.4}$$

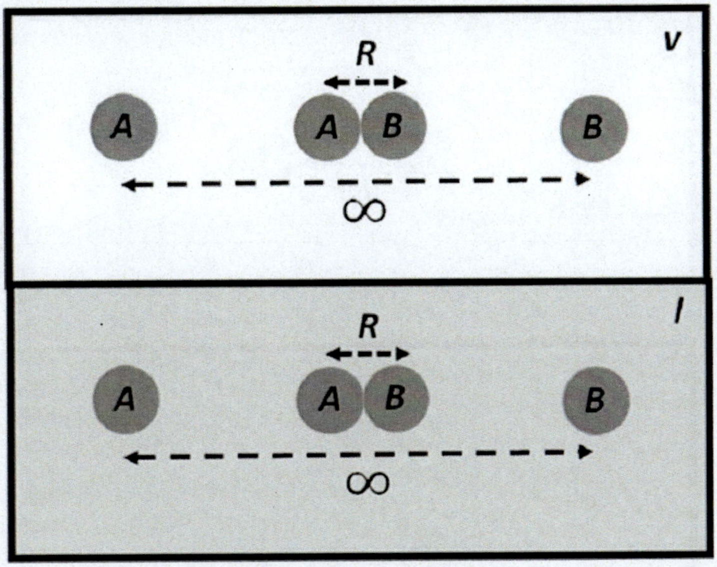

Fig. 1.5 The same process as in Fig. 1.1, but the process is carried out in both phases v and l. The SII is defined as the difference in the work associated with the same process, denoted ($\infty \to R$) in the two phases. See Eq. (1.4)

In Eq. (1.4), we deduct the work in a vacuum (v) from the work in the liquid phase (l). We assume that the *direct* interaction energy $U(R)$ is approximately the same in two phases. Therefore, the difference, defined in (1.4) is the *indirect* interaction or the SII. The argument R is used here and below instead of the more detailed description of the process ($\infty \to R$).

It should be noted that the assumption made above is not always valid. If the solvent molecules cause strong polarization effects, then the *direct* interaction between the two solute particles A and B might be modified. A simple example is shown in Fig. 1.6. The direct interaction between A and B in a vacuum $U(R_{AB})$, might be modified into $U'(R_{AB})$, in the presence of a polarizing particle C. In this book, we shall always ignore such polarization effects.

Thus, we write the work (either ΔA or ΔG) of the process ($\infty \to R$) in the liquid, as a *sum* of two contributions:

$$W^{(l)}(R) = U(R) + \delta W(R) \tag{1.5}$$

It should be noted that traditionally the SII was studied in the context of the interaction between two (or more) electrically charged particles, see Gurney [4]. The interaction energy between two-point charges ε_A and ε_B in vacuum is given by the well-known Coulomb's Law:

$$U(R) = K\frac{\varepsilon_A \varepsilon_B}{R} \tag{1.6}$$

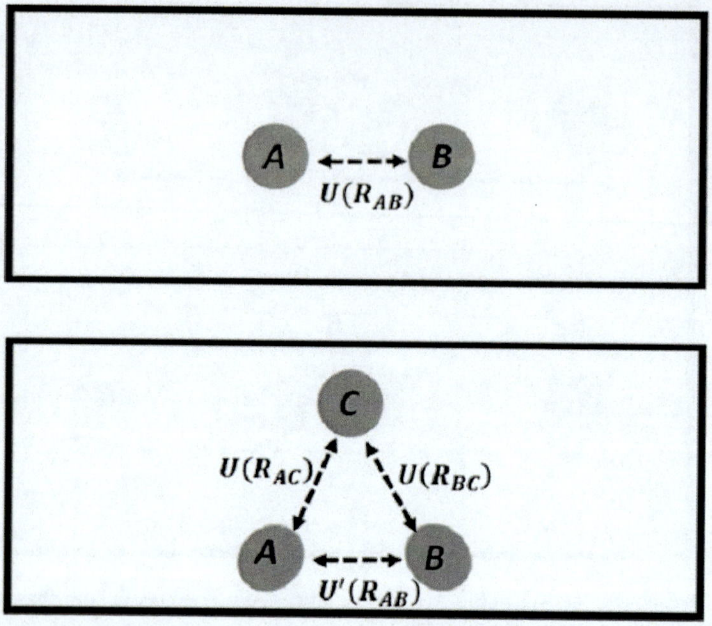

Fig. 1.6 The direct interaction $U(R_{AB})$ may be modified to $U'(R_{AB})$ in the presence of a third particle C, which polarizes A and B

where K is a constant which depends on the units we choose for the various quantities in the equation, and R is the distance between the two charges.

Here, we are interested only in the modification of the function $U(R)$ when we have the same two charges in a medium other than a vacuum. It is well known that when the two charges are in a solvent, say, liquid water, we write instead of Eq. (1.6) the modified equation:

$$W^{(l)}(R) = K\frac{\varepsilon_A \varepsilon_B}{RD} \tag{1.7}$$

where D is the macroscopic dielectric constant of the medium, say, water. This constant is taken from macroscopic measurements, say, on the interaction between two charged plates when they are inserted in a medium. One cannot expect that Eq. (1.7) will hold between two microscopic particles with the same constant D, independently of the distance between the two charges.

As we have seen above, for non-charged particles and with polarization effects, the modified work $W(R)$ is obtained from $U(R)$ by *adding* a term (the SII) not by multiplication by a constant (D). Statistical mechanics provides the means of systematically studying SII and its dependence on the temperature and density of the solvent.

1.2 The Connection Between the Work for the Process ($\infty \to$ R) and Solvation Gibbs or Helmholtz Energies

We refer the reader again to Fig. 1.5, where the same process (denoted ($\infty \to R$) is carried out in two phases; an ideal gas (which approximately may be assumed to be a vacuum), and a liquid phase. We now do a full cyclic process, see Fig. 1.7.

First, we do the process in a vacuum, as in Fig. 1.1, i.e. we bring A and B from fixed positions, but at infinite separation to some new configuration at which the distance between A and B is R. Next, we transfer the pair AB a fixed distance R into the liquid phase. If the system is at constant T and P, the corresponding work in this step is the Gibbs energy of solvation of the pair, AB which we denote by $\Delta G^*_{AB}(R)$. Next, we separate the two particles at a distance R and move them to an infinite separation, i.e. we do the "reverse of the process ($\infty \to R$). Finally, we transfer the two separate particles from the liquid to the gaseous phase. This part involves the de-solvation of the two particles, with the corresponding Gibbs energies $-\Delta G^*_A$ and$-\Delta G^*_B$ of A and B, respectively. Since we arrived at the initial state, the total change in Gibbs energy in the entire cyclical process must be zero. Hence, we write:

$$U(R) + \Delta G^*_{AB}(R) - W^{(l)}(R) - \Delta G^*_A - \Delta G^*_B = 0 \tag{1.8}$$

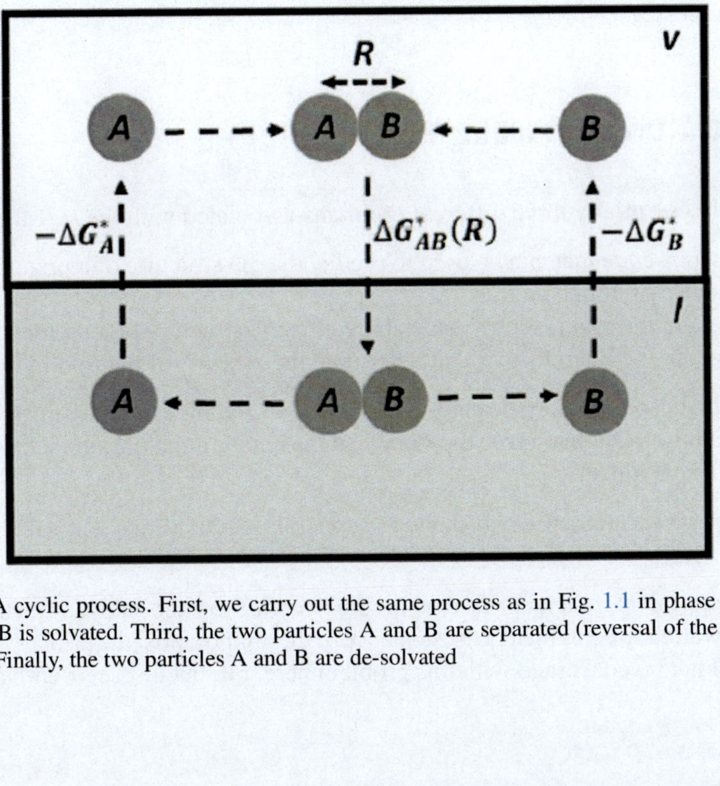

Fig. 1.7 A cyclic process. First, we carry out the same process as in Fig. 1.1 in phase v. Second, the pair AB is solvated. Third, the two particles A and B are separated (reversal of the process in Fig. 1.2). Finally, the two particles A and B are de-solvated

This equation may be rearranged to obtain the SII in terms of Gibbs energies of solvation:

$$\delta W(R) = W^{(l)}(R) - U(R) = \Delta G^*_{AB}(R) - \Delta G^*_A - \Delta G^*_B \tag{1.9}$$

Thus, the SII is simply the difference in the solvation Gibbs energy for the two configurations of the pair of particles; at a distance R, and at an infinite distance (in the latter configuration the solvation Gibbs energy of the pair is simply the sum of the solvation Gibbs energies of A and B, separately).

It should be noted that had we carried out the process ($\infty \to R$) at constant T and V, the Gibbs energies on the right-hand side of (1.9) would be replaced by the Helmholtz energies of solvation. As we explained in the previous section (and in more detail in Ref. [2]), the difference between the Gibbs and the Helmholtz energies of solvation is usually very small and may be ignored.

Also, we note that when we deal with non-spherical particles, Eqs. (1.8) and (1.9) should be modified to include both the distance and the orientations of the two particles involved in the process.

Statistical mechanics provides the tools for expressing the SII as well as the solvation quantities in terms of the intermolecular interactions among all the particles in the system (see Chap. 2). In the rest of this chapter, we discuss the motivation for studying the SII in the context of some important biochemical processes such as protein folding and self-assembly of proteins. In Chap. 6, we shall return to these problems and examine the extent of success of the theories of SII in understanding these processes.

1.3 The Protein Folding Problem (PFP)

There are essentially three different problems associated with the so-called PFP.

1. Is there a *code* that can be used to predict the 3D structure of a protein, given its sequence of amino acids?
2. What are the main factors that stabilize the 3D structure of a protein?
3. What are the main factors that determine the "speed" of the folding process?

The first problem is associated with Anfinsen's hypothesis sometimes referred to as the Anfinsen Dogma. From the studies of the renaturation of ribonuclease *in-vitro*, Anfinsen concluded:

> ... the native conformation is determined by the totality of the inter-atomic interactions, and hence by the amino acid sequence, in a given environment.

An equivalent statement is that the *information* on the native structure of the protein is already contained in the sequence of the amino acids (the primary structure). We shall not be concerned with this problem here. For details, see Ben-Naim [5, 6].

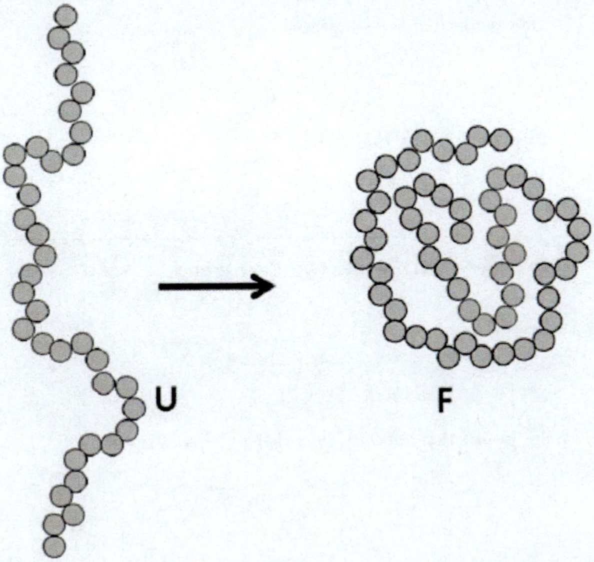

Fig. 1.8 A schematic process of folding of a protein U → F

The second problem led to the study of all the possible interactions among all the groups of the protein in the final 3D structure.

Consider the process of "folding" a short segment of a protein as shown in Fig. 1.8. The reaction may be written as:

$$U \rightarrow F \tag{1.10}$$

where U is the *unfolded* protein and F is the final *folded* protein. It is known that this process occurs spontaneously in aqueous solutions. Therefore, it is believed that the solvent-induced interactions (SII) are the main "driving forces" in this process (note that the term "driving forces" is commonly used for the thermodynamic Gibbs or Helmholtz energy change in this process. It is not a *force* as will be explained in Chap. 2 and in other parts of this book.

Consider the Gibbs energy change for the following hypothetical process. We start with a single unfolded protein at a fixed position and orientation (as well as fixed conformation such as the one shown on the left-hand side of Fig. 1.8.

Next, we fold the protein to obtain the conformation of the folded form on the right-hand side of Fig. 1.8, the process is carried out within the solvents T, P, and N constants.

The Gibbs energy change may be written as:

$$\Delta G^{l} = \Delta G(direct) + \Delta G(indirect) \tag{1.11}$$

where $\Delta G(direct)$ is the change of the Gibbs energy for this process in a vacuum, and $\Delta G(indirect)$ involves all solvent-induced effects in this process.

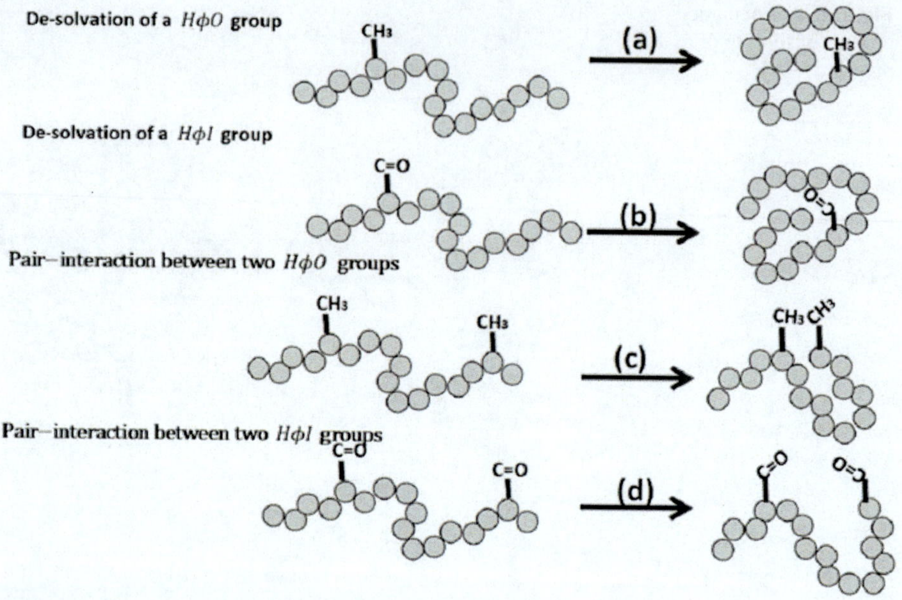

Fig. 1.9 Schematic description of four possible solvent-induced effects that were studied in connection with the protein folding process

In Fig. 1.9 we show four possible solvent-induced effects which occur in the process of protein fording:

(a) De-solvation of a $H\phi O$ group. A non-polar group such as a methyl group which was initially solvated by the solvent (water), is transferred into the interior of the protein where it is not solvated by water molecules.
(b) De-solvation of a $H\phi I$ group. A polar group such as an OH, NH or a CO group, which was initially solvated by water, is transferred into the interior of the protein.
(c) $H\phi O$ pair-interaction. This involves the transfer of two $H\phi O$ groups which are initially far apart from each other (hence, independently solvated). To some close distance R, at which the indirect, or solvent-induced interaction, is considerable.
(d) $H\phi I$ pair-interaction. This is the same as described in 3—but involves two $H\phi I$ groups.

Clearly, these four cases do not exhaust all possible solvent-induced effects. One can add to this list interactions among three or more groups which can either be all $H\phi O$, or $H\phi I$, or mixed types. It should be also emphasized that all these four effects involved groups that are attached to the protein. For a long time, the effects that were actually studied involved whole molecules, not groups of atoms that are part of the protein. The difference between these two kinds of effects is important and will be further discussed throughout this book. In the history of the study of SII in connection with the PFP the four effects involving molecules (not groups), were studied extensively by both experimental and theoretical means.

The third part of the PFP is associated with the name Levinthal [7, 8], and later appeared as one of the 125 "unsolved" ("What we don't know") questions in science, Kennedy and Norman [9]. It states:

> Out of a near infinitude of possible ways to fold, a protein picks one in just tens of microseconds. The same task takes 30 years of computer time.

This problem is sometimes referred to as the Levinthal Paradox. The truth is that there is no paradox at all. We shall further discuss this "paradox" in Sect. 6.2, for details, see Refs. [5, 10].

In the next section, we briefly survey the "evolution" of the "ideas" and "theories" regarding the main SII and forces.

1.4 Some Historical Landmarks in the Study SII in Attempting to Understand the PFP

In this section, we briefly survey the evolution of various ideas about the dominant factor (or factors) that could explain the PFP. We start with earlier ideas about the intramolecular HBs, the "natural" candidates which were believed to be responsible for the stability of proteins. Unfortunately, these "natural" factors which, were initially powerful enough to explain the stability of the protein, were later (wrongly) dismissed as insignificant. This led Kauzmann to introduce the so-called "*hydrophobic bond*"; an ill-defined and ill-understood idea that had dominated the field for over half a century- not because it offered a better explanation, but because it was the only one in store, and no one questioned its validity.

Only in the late 1980s and early 1990s, a reexamination of all possible SII was undertaken, which undermined most of the accepted ideas about the role of HBs and hydrophobic interactions. Instead, overwhelming evidence was found, supporting the idea that hydrophilic interactions (and forces) are probably the dominant factors in the PFP as well as in other important biochemical processes.

1.4.1 The Early Dominance of Hydrogen Bonds, and Its Unfortunate Fall

Until about 1955 intra-molecular hydrogen bonds (HB) reigned supreme in the field of proteins. As soon as biochemists developed methods for determining the structure of proteins, they realized that the two main features of the secondary structure of proteins—the helices and β-sheets—involved many internal hydrogen bonds between the carbonyl and the amino groups along the protein backbone. It was natural to conclude that hydrogen bonding is the main factor that contributes to the stability of the structure of the protein. The main proponent of this idea was Linus Pauling. One can follow the "evolution" of Pauling's ideas by comparing the successive editions of his book "The Nature of the Chemical Bond," Pauling [11, 12]. I believe that Pauling was correct in asserting that HBs are the major factor that contributes to the stability of the structure of both proteins and nucleic acids. Unfortunately, because of misinterpretation of some experiments carried out by Schellman, people looked for other factors for explaining the stability of proteins.

In 1939 Pauling summarized his views as follows:

> I believe that as the methods of structural chemistry are further applied to the physiological problems, it will be found that the significance of the hydrogen bond for physiology is greater than that of any other single structural feature.

The idea that HBs are important factors in maintaining the stability of proteins was natural and quite correct. Unfortunately, in 1955 this idea suffered a serious blow. This is a remarkable example of how a misinterpretation of an experimental result, by well-respected scientists had led to the discarding of a good and a correct idea, and paved the way for the acceptance of a dubious and an ill-understood idea.

The story starts in 1955, when John Schellman published a series of articles on the dimerization of urea in aqueous solutions. As is well known the urea molecules has two amine groups and one carbonyl group Fig. 1.10. The first group can serve as a "donor," while the second serves as an "acceptor" for HB, Fig. 1.11.

Schellman explained his experimental results by assuming that urea molecules form "dimers," held together by a HB between the amino group of one molecule and the carbonyl group of the second molecule. Thus, the deviations of aqueous urea

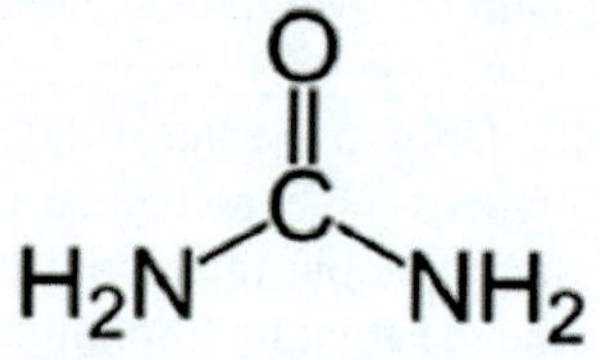

Fig. 1.10 The molecular structure of urea

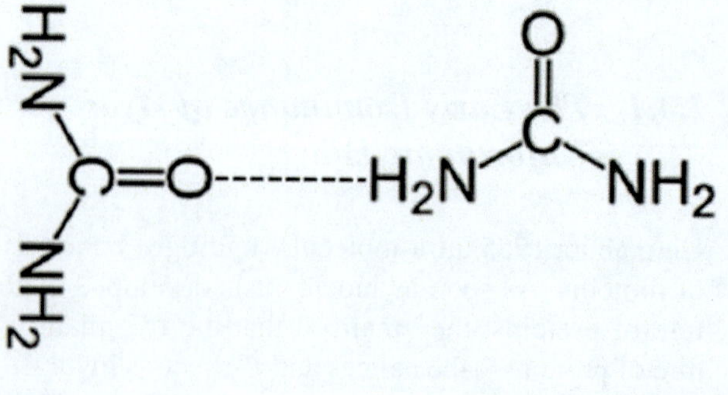

Fig. 1.11 A HB formed between the amine (the donor) and the carbonyl (the acceptor) groups of the urea molecule

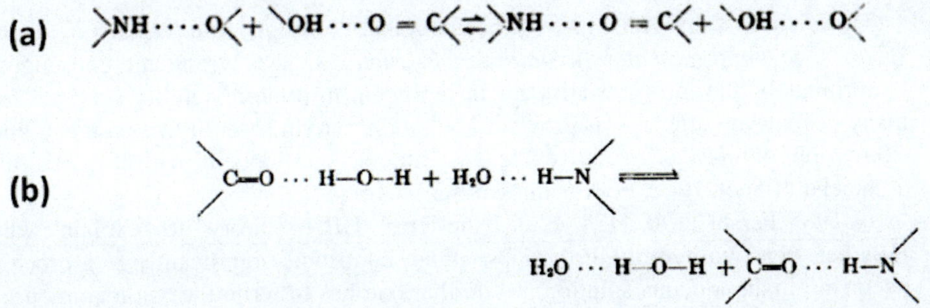

Fig. 1.12 **a** The stoichiometric reaction as written by Schellman [13, 14], and **b** as written by Kauzmann [15]

solutions from ideality was explained by the tendency of urea molecules to associate in water.

Schellman [13, 14], also wrote the "stoichiometric" reaction as shown in Fig. 1.12a, and concluded that:

> The peptide hydrogen bond, which is so important in modern theories of protein structure, involves groups which are almost identical with those of the urea molecules.

Clearly, Schellman believed that the energetics involved in the HB between two urea molecules is "almost identical" to the intramolecular HBs within the proteins. However, an error featuring in the "stoichiometric" reaction, that he wrote, see Fig. 1.12a, sowed the seeds for the future misinterpretation of his results. This error was propagated through the years by many well-known scientists.

In 1959, Kauzmann wrote a very influential review article [15] on: "Some factors in the interpretation of protein denaturation." In that review Kauzmann mentioned the work of Schellman, he first wrote the dimerization reaction as:

$$2\, urea(aq.) \rightleftarrows (urea)_2(aq.) \tag{1.12}$$

Then, he quoted the stoichiometric reaction, as written by Schellman, Fig. 1.12b. Referring to this reaction, Kauzmann concluded:

> From the rather small value of the f_u part of the formation of the hydrogen bond between urea molecules, J. A. Schellman concluded that 'hydrogen bonds, taken by themselves, give a marginal stability to ordered structures, which may be enhanced or disrupted by interactions of side-chains.

It should be noted that many years after the publication of Kauzmann's review [15], I visited John Schellman in Eugene, Oregon. I gave a talk on $H\phi O$ and $H\phi I$ interactions in proteins. I also mentioned his works from 1955 and Kauzmann's review. Schellman admitted that his "stoichiometric" reaction, as shown in Fig. 1.12a was misleading, but he denied ever concluding *"that "hydrogen bonds, taken by themselves, give a marginal stability to ordered structures."*

Thus, it seems that Kauzmann's statement that "*hydrogen bonds, taken by themselves, give a marginal stability to ordered structures,*" was wrong and unfounded. Unfortunately, the idea that HBs are insignificant in protein stability survived for many years, even after it was proven to be totally wrong [16]. In fact, you can find expressions of this idea in relatively recent textbooks on Biochemistry, see quotations at the end of Sect. 1.4.2. For details see Ben-Naim [2].

In 1985 Fersht et al. [17], coined the term "HB inventory" to "explain" why HBs between two hydrophilic groups cannot contribute significantly to a process occurring in an aqueous solution. To do that one has to write the "stoichiometric" reaction as follows:

$$E - \mathrm{OH} \cdots w + w \cdots \mathrm{O} = \mathrm{C} \rightleftarrows E - \mathrm{OH} \cdots \mathrm{O} = \mathrm{C} + w \cdots w \tag{1.13}$$

where E is an enzyme and S is a substrate. By "counting" the number of HBs "broken" on the left-hand side of the equation, and the number of HBS formed on the right-hand side of the equation, Fersht concluded:

> The inventory shows that the net change in the number of hydrogen bonds is zero.

Thus, Fersht essentially repeated the same wrong argument given by Kauzmann, but now this argument was given a *name*, which endows the wrong argument an instant respectability.

In 1991 I wrote an article on: "*On the role of hydrogen-bonds in protein folding and protein association,*" where I proved that this "HB inventory argument" is totally wrong, [16]. Some details of the proof will be discussed in Sect. 1.4.4, below. I sent a copy to both Kauzmann and Fersht. Kauzmann responded with expression of admiration, while Fersht never responded. Unfortunately, in spite of the existing proof, of the invalidity of Kauzmann's conclusion, that "HBs are insignificant," and Fersht's "HB inventory argument," textbooks continued to propagate the wrong ideas which originated from misinterpretation of Schellman's experimental results. In his book published in 1999, (eight years after my article, [16] was published), Fersht wrote [18]:

> A crude, but very effective, way of understanding the energetics of hydrogen bonding is to perform a hydrogen bond inventory, i.e., count the number and nature of hydrogen bonds on each side of the chemical equation. Suppose that – XH is a good donor and B a good receptor. Then, on each side of the equation, there are two good hydrogen bonds. There is then little change in enthalpy in the reaction. Suppose that – XH is a poor hydrogen bond donor, but B is a good acceptor. Then on the left there is one strong bond and one weak one. The same is true on the right. The same inventory can be performed if – XH is a good donor and B is a poor acceptor. Again, there should be little change in enthalpy during the reaction. Thus, hydrogen bonding should be relatively isoenthalpic.
>
> Although the hydrogen bond inventory is zero, hydrogen bonding is energetically favorable in the formation of enzyme–substrate complexes because of the increase in entropy on the release of bound water molecules.

Unfortunately, this seemingly "logical" argument is totally wrong. Similar conclusions appear even in recent textbooks, see end of Sect. 1.4.2. I shall explain why in Sect. 1.4.3 below.

1.4.2 *The Rise of the* HϕO *Effect*

The rise of the ***HϕO*** effects and its dominance in the field of proteins is attributed to Kauzmann's article published in 1959. However, some serious discussion of the SII effects in proteins already appeared in a conference in 1954: McElroy and Glass [19],

> A symposium on the mechanism of enzyme action.

In this symposium, Kirkwood gave a lecture on "*The nature of the forces between protein molecules in solution.*"

On page 6 we find:

> We shall therefore adopt the reasonable hypothesis that the forces of major importance are the electrostatic forces, and the exchange repulsive forces determining the size and shape of the protein molecules.

In the discussion, following Kirkwood's lecture, Eyring commented (page 16):

> It seems to me that you can have very large forces if you have oily patches that don't like to be exposed to the water and if two large molecules fit particularly well to the order of magnitude of whether a water molecule can go in or out.

To this comment Kirkwood answered:

> I think that one might say that the average attraction between the aliphatic group in a polar solvent is not so much due to their van der Waals attraction as to the fact that the displaced water molecules may interact more effectively with themselves when the aliphatic groups are in contact. I think that attention should be called to the role of water molecules which has been far too much oversimplified in electrostatic theory. It is certain that in order to get a clear quantitative picture it would be necessary to take into account the discrete structure of the water in the immediate neighborhood of the interacting protein molecules.

In this comment, Kirkwood clearly emphasizes that one should consider more seriously the effect of the solvent (here water molecules) in any discussion on the interactions between proteins. These are the SII interactions that have been "*too much oversimplified in electrostatic theory.*"

This was in 1954. Unfortunately, it took more than 30 years to implement Kirkwood's suggestion to take into account "*the discrete structure of the water in the immediate neighborhood of the interacting protein molecules.*"

At the same symposium, Kauzmann lectured on "*Denaturation of Protein and Enzymes.*"

On page 71 we find:

> The known types of intramolecular bonds which might be important in maintaining the native protein in its unique folded state are these:

1. *Hydrogen bonds*
2. *What might be termed hydrophobic bonds, referring to the tendency of the more hydrophobic amino acid residues (valine, leucine, phenylalanine, tryptophan, and possibly alanine and proline) to avoid the aqueous phase and adhere to one another in a sort of intramolecular "micelle" analogous to those occurring in soap and detergent solutions (64). In a sense, hydrophobic bonds also represent a kind of hydrogen bond (or "anti-hydrogen bond") because they arise by being unable to participate in the strong hydrogen bonding of the water molecules; that is, they are stabilized by the rather strong hydrogen bonds of water. They may therefore make a very large contribution to the stability of the native protein.*

   Hydrophobic bonds are the same as what many writers have called "van der Waals bonds," but this term does not seem to be appropriate because the forces ordinarily called van der Waals forces probably contribute only a small fraction to the total energy of hydrophobic bonds.
3. *Salt bridge. These are ionic bonds which might be expected to form between positively charged (lysine and arginine) and negatively charged (glutamic and aspartic acids) amino acid residues. Jacobsen and Linderstrøm-Lang (49) have, however, given strong reasons for doubting that these links are important in most proteins.*

Note carefully, that in 1954 Kauzmann still believed that the intramolecular HBs in proteins are important [19]. It is only after Schellman's experiments on the dimerization of urea in water that Kauzmann seemed to have changed his mind regarding the contribution of HBs to protein stability.

In 1959, Kauzmann reached the unfortunate conclusion, based on Schellman's experimental results, that:

> hydrogen bonds, taken by themselves, give a marginal stability to ordered structures.

If HBs do not contribute significantly to the stability of proteins, then what are the significant factors that contribute to the stability of the protein? This was the birth of the idea that $H\phi O$ might important. Kauzmann's idea was based on the fact that most of the $H\phi O$ aide chains of the protein are found in the interior of the 3D structure of the protein. He referred to this phenomenon as the "tendency of non-polar groups to adhere to each other." He also knew from the work of Frank and Evans, [20] that transferring a hydrophobic molecule from water into an organic liquid involves a large and negative Gibbs energy change. From these two facts, Kauzmann came up with a brilliant idea which he initially referred to as "$H\phi O$ bonding." He immediately speculated that:

> The hydrophobic bond is probably one of the more important factors involved in stabilizing the folded configuration in many native proteins.

It should be noted that the general idea of the role of the $H\phi O$ effect in the formation of micelles and membranes was long recognized. See Tanford [21], and Tanford and Reynolds [22]. However, for as long as HBs were recognized as the most important factors in stabilizing the structure of proteins, there was no need to invoke the $H\phi O$ effect in connection with proteins. It was only after the apparent "collapse" of the importance of the HBs ("HB, taken by themselves give marginal stability to ordered structure"), that Kauzmann suggested that "hydrophobic bonds" are also important in proteins.

Note also that Kauzmann was clearly cautious when he said that the $H\phi O$ bond is "probably one of the more important factors." He did not say that it is *the most* important factor, as so many others had concluded unjustly.

The qualitative description of the $H\phi O$-bond as the tendency of non-polar groups to "adhere to each other" was initially based on the data on the solvation Gibbs energies of non-polar molecules in water. Kauzmann argued that if the transfer of say, a methyl group being initially solvated by water into the interior of the protein would contribute − 1 to − 2 kcal/mol, , the same as the Gibbs energy for the transfer of methane from water into an organic liquid. If that is true, then it follows that for a protein with about 150 amino acids which has about 50 non-polar side chains there will be a very large and negative Gibbs energy change which will "drive" the folding of the protein into the compact 3D structure where nearly all the non-polar groups are found in its interior (hence, removed from the water because they "fear water").

This idea although far from being well understood captured the imagination of numerous scientists. In the early 1970s, I had many discussions with Kauzmann, the first being on his term "$H\phi O$-bond" which I suggested that he replace with *$H\phi O$ solvation*, and *$H\phi O$ interactions*, and the second, regarding the question about the relative importance of any of these effects on the stability of proteins. In fact, even the term $H\phi O$ interaction was interpreted as "reversal of stability" because adhering to each other was viewed as a reducing contact with the solvent.

Kauzmann agreed with me that the term "$H\phi O$ bond" was inappropriate. He expressed enthusiasm about my measure of the strength of the $H\phi O$ interaction between two non-polar molecules such as methane, ethane, and the like.

In 1980, I wrote my book on "Hydrophobic interactions," in which I summarized most of what had been known about $H\phi O$ solvation and $H\phi O$ interaction, [23]. Also, I described my own contributions to the field; a new definition and a new measure of the $H\phi O$ interactions. At that time, it was established that the SII between two (or more) non-polar solutes in water is far stronger than the SII between the same solutes in other organic solvents. However, I expressed some doubts regarding the relative importance of the various $H\phi O$ effects in the stability of proteins. In the preface of my book "Hydrophobic Interactions," I wrote:

> In spite of my researches in this field over almost 10 years, I cannot confirm that there is at present either theoretical or experimental evidence that unequivocally demonstrates the relative importance of the $H\varphi O$ interactions over other types of interactions in aqueous solutions.

My doubts were based on *lack of evidence* in favor of the contention that the $H\varphi O$ effect is the "*most important*" effect in the "driving force" for protein folding. How can one claim that one factor is more important, or most important when one does not have a full *inventory* of *all* the factors involved in protein folding?

In the 1980s most people in the field were convinced that the $H\phi O$ effects were the "most important" driving forces in protein folding. Kauzmann, who read and reviewed my manuscript agreed with me that there was solid evidence in favor of the contention that the $H\phi O$ effects are the *most* important in protein folding, but he still believed that it was very probable that they were important. His belief was not based on evidence that $H\phi O$ effects are important in protein stability, but on the lack of any other factors that could explain the stability of proteins.

Thus, for over 30 years after the publication of Kauzmann's article in 1959, most people believed that the $H\phi O$ effects are the most important in the stability of proteins. This was expressed in numerous articles culminating in an article in 1990 by K. Dill titled "Dominant forces in protein folding."

Similar conclusions have appeared in many articles and textbooks:

Whitford [24]:

> The importance of the hydrophobic effect was for a long time underestimated. Charged interactions and hydrogen bonds are not strong intramolecular forces because water molecules compete significantly with these effects.

Delvin [25]:

> An important non-covalent force that causes a polypeptide to fold into its native conformation are the hydrophobic interaction forces…Hydrogen bonds…although they contribute to the thermodynamic stability of protein's conformation, their formation may not be a major driving force for folding.

Voet et al. [26]:

> The hydrophobic effect, which causes non-polar substances to minimize their contacts with water is, the major determinant of native protein structure…Perhaps surprisingly, hydrogen bonds which are central features of protein's structures, make only minor contributions to protein stability.

All these statements were found wrong and misleading once new SII were discovered in the early 1990s.

1.4.3 The Cracks, and the Eventual Fall of the HϕO Effects in Protein Folding

I mentioned above my doubts regarding the importance of the various $H\phi O$ effects in protein folding. In the 1980s I had questioned many of the statements regarding the "dominance" of the $H\phi O$ effect. First, I found that the HBs were not "insignificant," as most people thought. Second, I found that Kauzmann's model was not appropriate for proteins (briefly, the solvation Gibbs energy of say, methane, was found to be very different from the solvation Gibbs energy of a methyl group attached to the protein which led me to study the "*conditional*" solvation Gibbs energies of non-polar groups, see Fig. 1.13). By the late 1980s, I embarked on a systematic study of all possible SII among groups attached to the protein. Clearly, solvation Gibbs energies of non-polar molecules did not feature in the inventory of those SII. Also, SII between *pairs* of *molecules* did not feature in the list. Instead, the SII between groups such as methyl, propyl, and the like. Finally, I found that new SII between

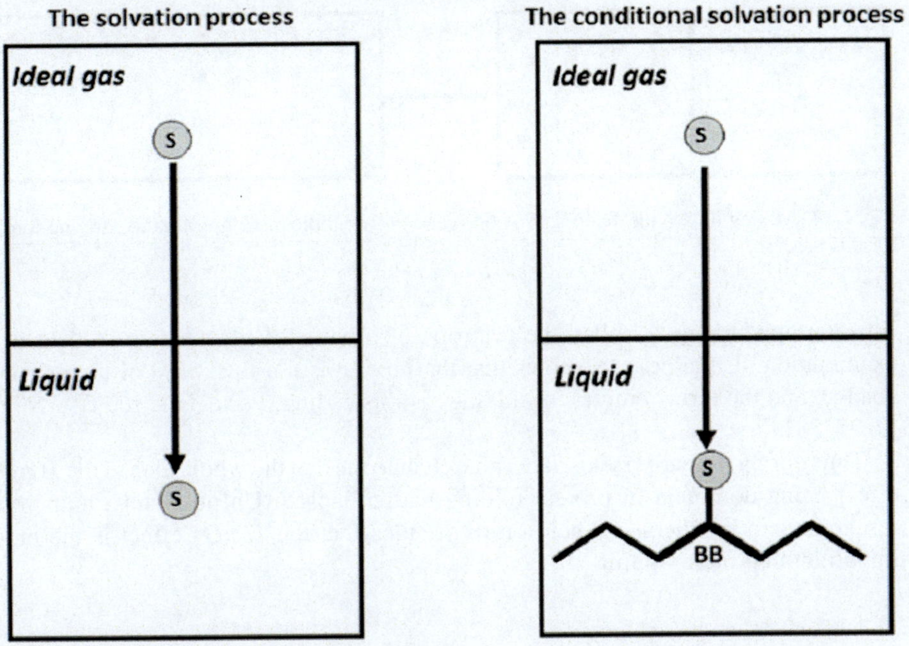

Fig. 1.13 The process of *solvation* of a solute S (left), and *conditional solvation* of the same solute near a backbone (BB) (right)

hydrophilic ($H\phi I$) groups might be far more important than the corresponding SII between $H\phi O$ groups.

The fact that non-polar groups on a protein, initially being *solvated* by water, and eventually are found in the interior of the protein, was the basis on which Kauzmann suggested the idea that $H\phi O$-bonds could be an important factor in stabilizing the structure of proteins. What Kauzmann did not realize was the fact that the Gibbs energy associated with the transfer of say, a methyl group from water to the interior of the protein is very different from the Gibbs energy associated with the transfer of methane from water into an organic liquid. This means that the entire model suggested by Kauzmann was inappropriate.

In addition, one can find in many articles and text books an argument based on the fact that $H\phi O$ groups are found in the interior of the proteins. This fact was turned into an argument, i.e., that the $H\phi O$ effect is also the "driving force" for the folding of proteins. Unfortunately, this argument is, in general invalid. Indeed, we observe that most $H\phi O$ groups are in the interior of the proteins, but this fact does not tell us anything about the "driving force," or the cause for the folding. This kind of argument, reminds me of a similar case of confusing an observable fact with its "driving force." This is the well-known experiment fact of mixing two ideal gases in an isolated system, Fig. 1.14. In this process, we observe mixing. Removing the partition between the two compartments leads to the mixing of the gases. One can easily calculate the entropy change for this process and find that it is always positive.

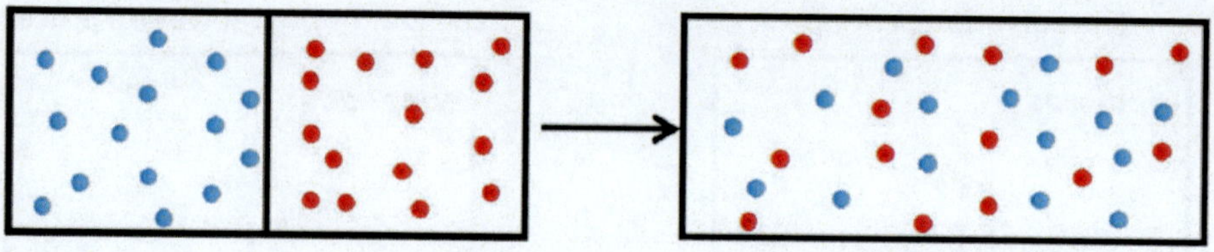

Fig. 1.14 Mixing of two ideal gases. We observe *mixing,* but the mixing is not the "driving force" for this process

This entropy change is called the "entropy of mixing." Unfortunately, a thorough examination of this process shows that the mixing is not the *cause* of the entropy change, and the term "entropy of mixing" is misleading. For details see Ben-Naim [2, 27, 28].

Thus during the early 1990s, it became clear to me that the whole idea of the $H\phi O$-effect being dominant in protein folding had collapsed. Unfortunately, numerous researchers in biochemistry held on to the idea that the $H\phi O$- effect is the most important in protein folding.

1.4.4 The Rise and Possible Dominance of the HϕI *Effects*

As a consequence of my detailed analysis of SII protein folding, I created some kind of inventory of all possible SII in protein folding, and later in any process occurring in aqueous media. In this inventory, there were SII between $H\phi O$ groups, between $H\phi I$ group, and between $H\phi O$ and $H\phi I$ groups. This inventory did not provide numerical estimates of the various SII. However, the first indication that SII involving $H\phi I$ groups might be important came from two sources:

1. Direct HBs between two $H\phi I$ groups had a significant Gibbs energy change (not insignificant as concluded by Kauzmann and many others).
2. Some experimental results indicated that SII between two $H\phi I$ groups might be very strong, perhaps even stronger than the corresponding interaction between two $H\phi O$ groups.

Recall that the birth of the $H\phi O$ was followed from Kauzmann's conclusion about the insignificant contribution of HBs. In 1991 I published an article titled: "*On the role of hydrogen bonds in protein folding and protein* association." In this article, I showed that HBs in proteins are not negligible. The main reasons for my conclusions are the following:

First, on the left-hand side of the stoichiometric reaction, Eq. (1.13), the two $H\phi I$ groups are not HBed to water, but rather solvated by water molecules. Therefore, the "counting" is not of the number of HBs on the two sides of the equation but one HB is formed on the right-hand side of the equation, and two solvation energy (or enthalpy, or Gibbs energy) on the left-hand side. Thus, the very writing of the stoichiometric equation in the form of Eq. (1.13) is misleading.

Second, the term "w ··· w" on the right-hand side of the equation should not appear in the stoichiometric equation (as written originally by Schellman, Kauzmann, Fersht, and many others). Instead, one should write the stoichiometric reaction as:

$$[\mathrm{E-OH}]_{\mathrm{solv.}} + [\mathrm{S-C=O}]_{\mathrm{solv.}} \rightleftarrows [\mathrm{E-OH\cdots O=C-S}]_{\mathrm{solv.}} \tag{1.14}$$

The reason is that both the enzyme and the substrate are solvated (solv.) by water molecules in both the initial and the final states. In this reaction, the water molecules released from the "solvation sphere" of the enzyme and of the substrate flow into the bulk of the water at a constant chemical potential. Therefore, whatever HBs are formed by the water molecules should not be considered in this "HB inventory." More details on this in Ben-Naim [6, 16]. Later, I found many more $H\phi I$ effects involving more than two $H\phi I$ groups which I will describe in the next chapters of this book. Here, I want to give a simple intuitive explanation of the two $H\phi I$ effects mentioned above based on a very simple model for water (see Appendix B).

Suppose that we view a water molecule as a Lennard–Jones sphere with four "arms" emanating from its center. The LJ sphere provides a strong repulsive interaction between two molecules when they approach a distance smaller than, say, 2.8 Å. The four "arms" are the directions along which a water molecule can form HB with other water molecules, or any group that can from HBs. In Appendix A and B we note that one should distinguish between two "donor" arms and two "acceptor" arms (which are approximately in the direction of the OH group and the lone-pair in a water molecule). Here, we assume, for simplicity that all the arms are equivalent.

The qualitative reasons for expecting the two $H\phi I$ effects mentioned above are:

1. Direct HB at a distance of R = 2.76 Å

Figure 1.15 shows two water molecules brought from fixed position and orientation but at infinite separation to a final configuration at which a direct HB is formed. The whole process is carried out in water.

In Appendix A one can find estimates of the solvation Gibbs energy of a water molecule in liquid water. Also, an estimate of the solvation Gibbs energy per one arm of a water molecule in liquid water. We now calculate the change in the number of solvated arms in the process shown in Fig. 1.15. The solvation Gibbs energy per arm was estimated at − 2.25 kcal/mol. In the initial state all the eight arms of the two water molecules are solvated. Hence, the

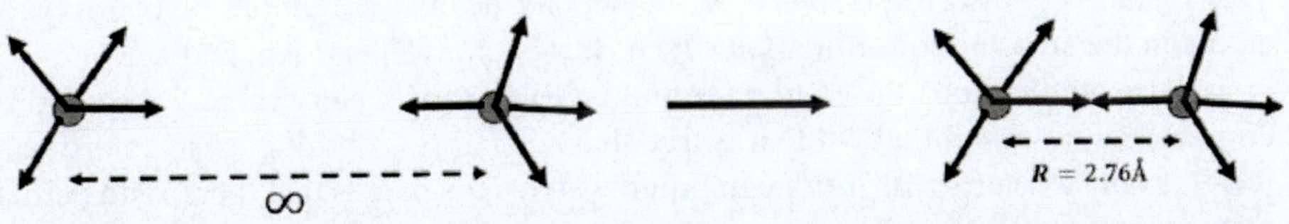

Fig. 1.15 Two water molecules form a direct HB at a distance of about R = 2.76 Å

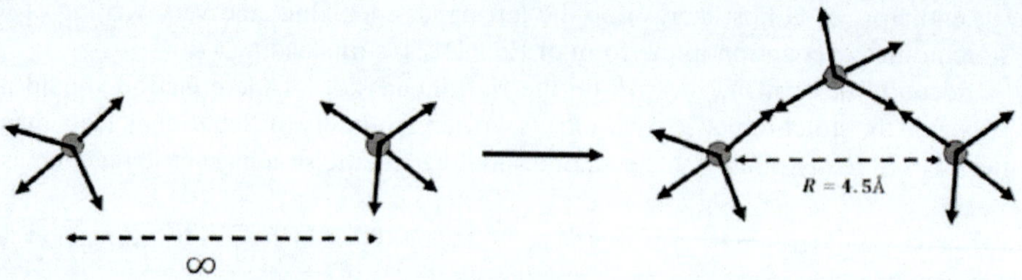

Fig. 1.16 Two water molecules at a distance of R = 4.5 Å may be connected by a HB-bridge

total Gibbs energy of solvation is about $-8 \times 2.25 = -18$ kcal/mol. In the final configuration, two of the arms are engaged in a direct HB. Therefore, only six arms are solvated, i.e. the "dimer" on the right-hand side in Fig. 1.15 has a solvation Gibbs energy of about $-6 \times 2.25 = -13$ kcal/mol. Hence, the solvation Gibbs energy in this process changes by the amount $+$ 4.5 kcal/mol. Taking the HB energy in the process to be about -6 kcal/mol, we find that the net change in the Gibbs energy is about -1.5 kcal/mol. This is certainly not a negligible contribution as was estimated by Kauzmann, Fersht, and others.

2. Indirect interaction at a distance R = 4.5 Å

Figure 1.16 shows a process of bringing two water molecules from a fixed configuration and at infinite separation to the final configuration at which the two water molecules are at a distance of 4.5 Å, and are oriented in such a way that a water molecule can form two HBs with these two water molecules.

Here again, we can use the simple "counting" of the number of solvated arms in the initial and the final states. However, in the final states the solvation Gibbs energy of two arms by a water molecule is different from the sum of the solvation Gibbs energy of two independent arms. The total solvation Gibbs energy change for this process was first estimated by Ben-Naim in the early nineties, and is also discussed in Appendix B. To my surprise, the Gibbs energy change for the process in Fig. 1.16 was found to be larger for the process in Fig. 1.15, although the former does not involve direct HBs between the two water molecules.

Soon after the publication of this new *HϕI* interaction, I found some experimental data in the literature which supported my estimates, this is discussed in Chap. 5 of this book. Also, the idea that strong SII exists at a distance of R = 4.5 Å was consistent with the well-known second peak in the radial distribution function of liquid water. This distance is also the distance of the second nearest neighbors of two oxygen atoms in the structure of ordinary ice (see Ben-Naim [29] and Appendix A).

In later publications I used to refer to this SII due to "water bridge." It should be emphasized however, that while it is true that such a water bridge exists in ordinary ice, it is only a "statistical" bridge in liquid water. This means that there is no permanent bridge connecting the two water molecules. Instead, we can say that the two water molecules have a high probability of formation of such a bridge between two water molecules at this specific configuration. The statistical character of this bridge was confirmed by many simulated calculations which are discussed in Chap. 5.

Once these two $H\phi I$ interactions were discovered and were found to be far stronger than the corresponding $H\phi O$ effects, it became clear to me that the dominance of the $H\phi O$ effects in protein folding would soon collapse. Unfortunately, the process of "paradigm shift" in science is quite slow. People continue to adhere to the unfounded myth that the $H\phi O$ effect is the most important factor in protein chemistry, ignoring the overwhelming evidence in favor of the $H\phi I$ effects. Sometimes, when I give a talk on $H\phi O - H\phi I$ and emphasize the relative strength of the $H\phi I$ effect, someone in the audience raises the question: Accepting that each single $H\phi I$ effect is weaker than the corresponding $H\phi O$ effect, how do you know that the combined effects of all the $H\phi O$ groups is not larger than the combined effects of all $H\phi I$ groups?

My answer is very simple. In a typical protein of about 150 amino acids, there are about 50 (i.e. 30%) of $H\phi O$ groups. Let us assume that there are also about 50 $H\phi I$ side-chains. In addition, the backbone of such a protein contains another 300 of $H\phi I$ groups. It is clear the totality of the $H\phi I$ effects is far larger than that of the $H\phi O$ effects.

1.5 Some Concluding Remarks

In this chapter we surveyed the motivations for studying SII in connection with the process of protein folding and protein stability in aqueous solutions. The study of SII is not an easy one, particularly when the solvent is water. This fact led to all kinds of speculations about the main factors which are responsible for the stability of the structure of proteins, on one hand, and about the forces that are responsible for the speed of the folding of proteins. Until the present days, there are numerous scientists who cling to the view that the $H\phi O$ effects are the most important ones in determining both the stability and the speed of folding proteins. These views are changing albeit slowly in favor of the various $H\phi I$ effects.

Interestingly, the origin of the study of SII was associated with the problem of protein folding. However, once stronger $H\phi I$ interactions and forces were discovered they could be used to explain not only the stability of proteins and the speed of the folding process but other biochemical processes such as protein–protein association, molecular recognition and protein solubility in water. We shall revert to these problems in the final Chap. 6 of this book.

References

1. Ben-Naim A (1992) Statistical thermodynamics for chemists and biochemists. Plenum Press, New York
2. Ben-Naim A (2006) Molecular theory of solutions. Oxford University Press, Oxford, UK

3. Ben-Naim A (2017) Statistical thermodynamics, with applications to life sciences. World Scientific, Singapore
4. Gurney RW (1953) Ionic processes in solutions. Dover, New York
5. Ben-Naim A (2013) The protein folding problem and its solutions. World Scientific, Singapore
6. Ben-Naim A (2016) Myths and verities in folding theories. World Scientific, Singapore
7. Levinthal C (1968) Are there pathways for protein folding. J Chim Phys 65:44–45
8. Levinthal C (1969) Mossbauer spectroscopy in biological systems: Proceeding of a meeting held at Allerton House, Monticello, Illinois. In: De Brunner JTP, Munck E (eds.) Chicago, Illinois, University of Illinois Press, pp 22–24
9. Kennedy D, Norman C (2005) What don't we know? Science 309, No. 5731, 75
10. Ben-Naim A (2012) Levinthal's question revisited, and answered. J Biomol Struct Dyn 30:113–124
11. Pauling L (1939) The nature of chemical bond, 2nd edn. Cornell University Press, Ithaca, New York (1948)
12. Pauling L (1960) The nature of chemical bond, 3rd edn. Cornell University Press, Ithaca, New York
13. Schellman JA (1955) The thermodynamics of urea solutions and the heat of Formation of the peptide hydrogen bond. Comptes rendus des travaux du laboratoire Carlsberg, Serie Chimique 29:223–230
14. Schellman JA (1955) The stability of hydrogen-bonded peptide structures in Aqueous solution. Comptes rendus des travaux du laboratoire Carlsberg, Serie Chimique 29:230–259
15. Kauzmann W (1959) Some factors in the interpretation of protein denaturation. Adv Protein Chem 14:1–63
16. Ben-Naim A (1991) On the role of hydrogen-bonds in protein folding and protein association. J Phys Chem 95:1437–1444
17. Fersht AR, Shi JP, KmilI-Jones J, Lowe DM, Wilkinson AJ, BIow DM, Brick P, Carter P, Waye M, Winter G (1985) Nature 314:235–238
18. Fersht A (1999) Structure and mechanism in protein science. W. H. Freeman and Comp, New York
19. McElroy WD, Glass B (1954) A symposium on the mechanism of enzyme action. In: McElroy WD, Glass B (eds) Johns Hopkins Press, Baltimore
20. Frank HS, Evans MW (1945) J Chem Phys 13:504
21. Tanford C (1973) The hydrophobic effect: formation of micelles and biological membranes. Wiley-Interscience, New York
22. Tanford C, Reynolds J (2001) Nature's robots, a history of proteins. Oxford University Press, Oxford
23. Ben-Naim A (1980) Hydrophobic interactions. Plenum Press, New York
24. Whitford D (2005) Proteins: structure and function. John Wiley, and Sons, LTD., New Jersey
25. Delvin MT (2006) Textbook of biochemistry with clinical correlations, 6th edn. Wiley-Liss, Hoboken, NJ
26. Voet DJ, Voet JG, Pratt CW (2008) Principles of biochemistry, 3rd edn. John Wiley and Sons, Hoboken, NJ
27. Ben-Naim A (1987) Solvation thermodynamics. Plenum Press, New York
28. Ben-Naim A (2006) Entropy of mixing and entropy of assimilation, an information-theoretical prospective. Am J Phys 74:1126–1135
29. Ben-Naim A (2009) Molecular Theory of Water and Aqueous Solutions, Part I: Understanding Water. World Scientific, Singapore

Chapter 2
A Brief Introduction to Statistical Thermodynamics

In this chapter, we present a very brief introduction to some of the fundamental concepts of statistical thermodynamics (ST) that we shall need in this book. We shall not derive any relationship, nor prove any statement. The main purpose of this chapter is to provide a brief summary of the basic concepts of Statistical Mechanics and to establish the notations that will be used in the rest of the book. We also introduce the concept of molecular distribution function, which will be used in the definition of the SII. For more details, the reader is referred to Ben-Naim [1, 2].

2.1 Statistical Thermodynamics and Thermodynamics

The reader is supposed to be familiar with the basic theory of thermodynamics and some elementary concepts of Statistical thermodynamics (ST) of systems at equilibrium. Thermodynamics deals mainly with experimentally measurable quantities associated with macroscopic systems, and the relationships between these quantities. Examples are entropy, volume, temperature, and pressure. ST aims at calculating, either exactly or approximately, the same *macroscopic* quantities, from the knowledge of the properties of the individual particles and their interactions among them. We always assume that the macroscopic system consists of a huge number of particles and that the system is at equilibrium.

While thermodynamics deals with relationships between thermodynamic quantities, ST provides the thermodynamic functions themselves, i.e. the *explicit functional* dependence of, say, the entropy of a system on temperature, volume and number of particles in the system. The most famous example is the *entropy function* of an ideal gas:

$$S(T, V, N; m) = \frac{5}{2}kN - kN \ln \rho\Lambda^3 \tag{2.1}$$

A. Ben-Naim, *Solvent-Induced Interactions and Forces in Protein Folding*, Physical Chemistry in Action, https://doi.org/10.1007/978-3-031-31871-9_2

where $\rho = N/V$ is the total number density, k is the Boltzmann constant, and Λ^3 is the momentum partition function (or the de Broglie wave length) defined by:

$$\Lambda^3 = \left(\frac{h}{\sqrt{2\pi mkT}}\right)^3 \tag{2.2}$$

Here, h is the Planck constant and m is the mass of the particles. In Ben-Naim [3] one can find an interpretation of this expression for the entropy in terms of Shannon's measure of information (SMI). We shall not need that in this book.

2.2 The Two Postulates of Statistical Thermodynamics

ST start with the assertion of two postulates, on which the entire theory of ST has been developed. Postulates, like axioms in mathematics, are statements that everyone, or at least those who want to use ST, have to accept. As we shall soon see, these postulates are reasonable, but they are far from being trivial.

The First Postulate
The time average of a quantity measured on a single macroscopic system is equal to the ensemble average of the same quantity.

The meaning of this postulate is quite simple to understand. Let us discuss a simple example. Consider a single macroscopic system of, say, a one-component gas at a fixed pressure P and temperature T. The fixed pressure is maintained by the atmospheric pressure on the moving piston, see Fig. 2.1. Clearly, the height of the piston is proportional to the volume of the system.

We observe that the piston, hence the volume, of the system changes with time. These "fluctuations" in the volume are easy to understand. At any given moment the molecules of the system collide with the piston. These collisions produce the

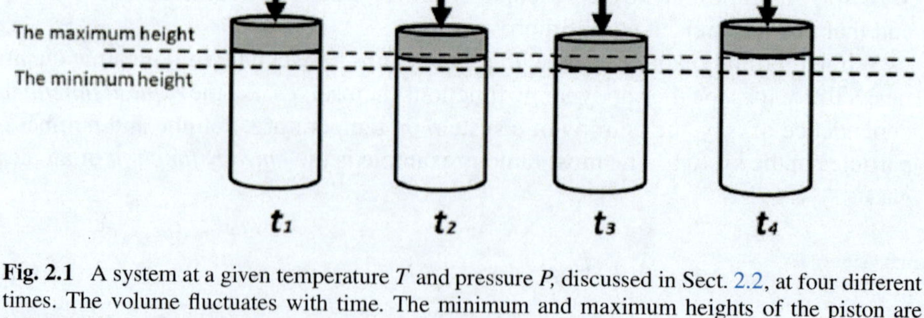

Fig. 2.1 A system at a given temperature T and pressure P, discussed in Sect. 2.2, at four different times. The volume fluctuates with time. The minimum and maximum heights of the piston are indicated by the dashed lines

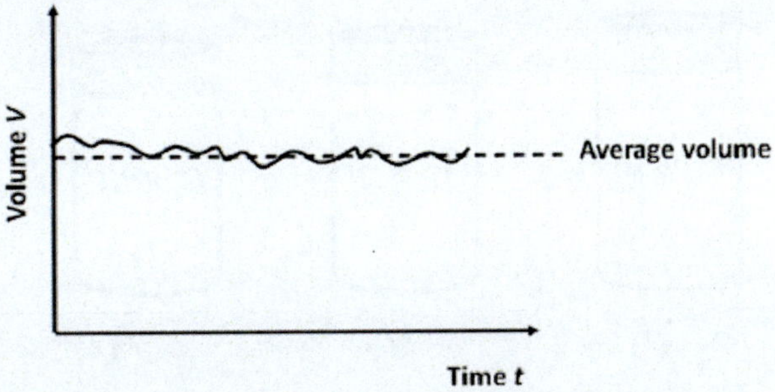

Fig. 2.2 The volume of the system changes as a function of time. The average volume is indicated by the dashed line

measurable pressure in the system. Clearly, the number of molecules and their kinetic energies changes with time. Hence also the pressure on the piston will change with time. Now, suppose that we could measure the volume of the system continuously as a function of time. We can denote this function by *V(t)*. See Fig. 2.2.

The volume that we actually measure at equilibrium is an average volume during the time of measurement. The time average (ta) of the volume is defined by:

$$\langle V \rangle_{ta} = \frac{1}{\Delta t} \int_{0}^{\Delta t} V(t) dt \tag{2.3}$$

Here we recognize that each measurement takes a finite interval of time, say $\Delta t \approx 1$ s. Therefore, what we actually measure is not the instantaneous *value* of the volume at a fixed *point-time, t*, but an average value of the volume—averaged over the time interval Δt. We denoted this time average by $\langle V \rangle_{ta}$.

Unfortunately, we cannot either calculate or measure the actual function $V(t)$, and then calculate the average over the time interval Δt. To overcome this problem, Gibbs suggested the following idea. Suppose that instead of measuring the volume of a *single* system at each point-time *t*, we envisage an *ensemble* of systems. The ensemble contains a very large number of thermodynamically identical systems M, i.e. each system is characterized by the same independent variables, say, T, P, N, Fig. 2.3.

Let V_i be the volume of the *i*-th system in the ensemble. The ensemble average volume is defined by:

$$\langle V \rangle_{ea} = \frac{1}{M} \sum_{i} V_i \tag{2.4}$$

The *first postulate* states that the *time average* (Eq. 2.3) of the volume of a single system ($\langle V \rangle_{ta}$) will be equal to the *ensemble* average, Eq. (2.4), ($\langle V \rangle_{ea}$).

$$\langle V \rangle_{ta} = \langle V \rangle_{ea} \tag{2.5}$$

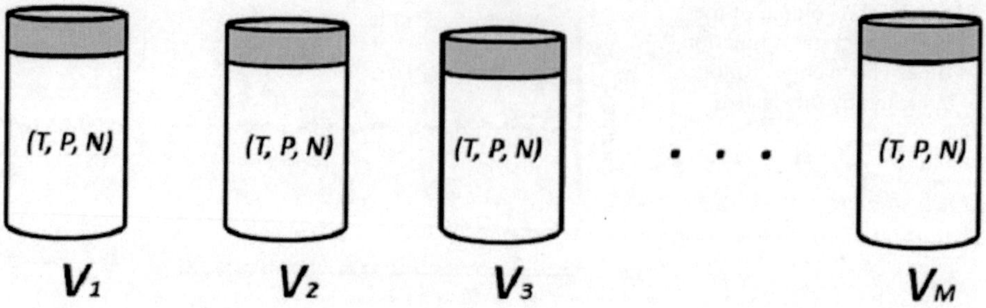

Fig. 2.3 An ensemble of M systems all at the same, *T, P, N*. The volumes of the systems might be slightly different for each individual system

A simple way of understanding this last equation, consider the following example. We throw a die n times and obtain a sequence of results at each point in time. If the die is fair, the average outcomes will be about:

$$\text{Average} = \frac{\sum n_i}{n} \approx 3.5 \tag{2.6}$$

On the other hand, one can throw M dice at once (an ensemble of dice) and calculate the average outcome of the ensemble of dice. Assuming that all the dice are identical, we expect that the average over the sequence (in time) to be equal to the average over the ensemble.

The Second Postulate

In an isolated system, having a fixed volume energy and composition at equilibrium, all possible microstates are equally probable.

Clearly, this postulate cannot be proved. However, it is plausible if we do not know any additional information on the system. Think again of a fair die. If the die is *known* to be fair, then by definition of a fair die, all of the six outcomes have the same probability. However, if we do not know anything about the die, the *best guess* we can make is to assume that the die is fair and that each outcome has the same probability of occurrence. For a macroscopic isolated system, the postulate states that all the quantum mechanical states have equal probability.

On these two postulates, the whole edifice of statistical thermodynamics may be constructed by mathematical reasoning. It should be noted that in the old literature on ST another postulate featured; that a macroscopic system consists of a very large number of atoms or molecules. This postulate is nowadays considered as a fact not a postulate.

2.3 The Boltzmann Definition of Entropy

Suppose that an isolated system, characterized by the energy *E*, the volume *V* and the number of particles *N*, there are altogether Ω microstates, i.e. Ω quantum mechanical states. The *Boltzmann entropy* is defined by:

$$S_B = k \ln \Omega \tag{2.7}$$

However, since Clausius had already defined entropy within thermodynamics, we view this as a new *definition* of entropy. {in fact Clausius did not define entropy itself, but a small change in entropy for a specific process (see Ref. [4])}. Clearly, one must show that the Boltzmann definition is equivalent to Clausius'. This has not been done in general, i.e. there is no formal proof that the Boltzmann definition is identical to the Clausius definition. However, for many systems, for which the entropy may be calculated, one finds agreement between the Boltzmann entropy and the Clausius entropy. Therefore, we accept these two definitions as being equivalent. It should be emphasized that this definition of entropy is for an isolated system, i.e. a system characterized by the variables (E, V, N). Where *N* is either the number of identical particles, or the vector $\boldsymbol{N}$ describing the composition of the system, $\boldsymbol{N} = (N_1, N_2, \ldots, N_c)$.

In Eq. (2.7), $k = 1.38 \times 10^{23}\ \mathrm{JK}^{-1}$ is called the *Boltzmann constant*, and Ω is the total is the number of microstates.

Since all the thermodynamic quantities derived from Boltzmann's entropy behave the same way as the Clausius' entropy, we shall drop the subscript *B* in S_B, and simply use the letter *S* for the *thermodynamic entropy.*

It should be noted that since Ω in Eq. (2.7) is the *total number of microstates*, the entropy is not defined for every single microstate, nor for any subset of microstates. It follows that the Boltzmann entropy is defined for systems at equilibrium, at which all the microstates are accessible. It should be mentioned here that there is another definition of entropy which is *different* from both Clausius' and Boltzmann's definitions. This definition is based on Shannon's measure of information. It can be shown that this definition is also equivalent to the two definitions mentioned above. It is equivalent in the same sense that Boltzmann's definition is equivalent to Clausius'. We shall not need this definition in the present. The interested reader is referred to Ben-Naim [3, 4] for details.

2.4 Ensembles, and the Fundamental Thermodynamic Quantities

In this section, we present a few useful ensembles in statistical mechanics.

2.4.1 *The Isolated System; The* E, V, N *Ensemble*

The starting ensemble in ST is one consisting of M isolated systems, Fig. 2.4. An isolated system is defined as a system having a fixed energy E, a fixed volume V, and a fixed number of particles N. When dealing with multi-component systems, N is replaced by the vector $\boldsymbol{N} = (N_1, N_2, \ldots, N_c)$, where N_i is the number of particles of the i-th kind. Although a truly isolated system is of no interest in actual experimental work, an isolated system, and the corresponding ensemble are very useful in the foundations and development of the theory of ST.

Let Ω be the number of quantum mechanical states of the isolated system characterized by the variables E, V, N. These are all the eigenstates of the energy E, i.e., the degeneracy of the solution of Schrödinger equation for this system. In classical statistical thermodynamics, we describe the microstates of a system, characterized by E, V, and N, by giving the locations and the momenta of all the particle. This is true for a system of simple particles. For more complex particles we need more parameters such as angles of orientation of each molecule as well as angle of internal rotations within the molecule, to describe the microstate of the system.

A micro-state of such a system is a list of all the locations and all the momenta of all the N particles; this is a huge vector $\left(\boldsymbol{R}_1, \boldsymbol{R}_2, \ldots, \boldsymbol{R}_N, \boldsymbol{p}_1, \boldsymbol{p}_2, \ldots, \boldsymbol{p}_N\right)$, where each of the $\boldsymbol{R}_i$ and each of the $\boldsymbol{p}_i$ has three components (x_i, y_i, z_i) and $\left(p_{ix}, p_{iy}, p_{iz}\right)$. Altogether, this vector has 6N components. For a strictly classical system, there is an

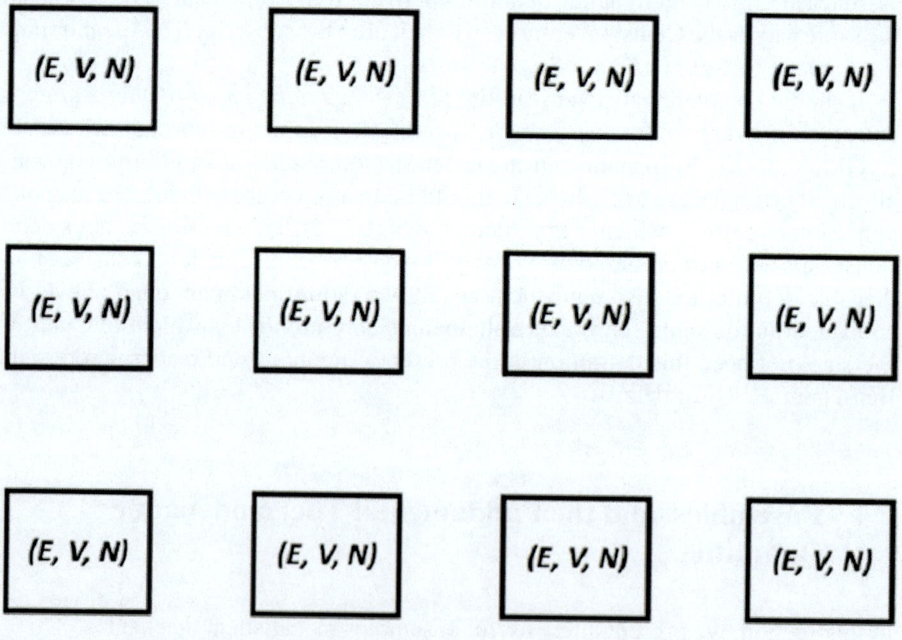

Fig. 2.4 An ensemble of isolated systems. Each system is characterized by the same set of variables E, V and N

infinite number of micro-states. This is so because each of the variables $\boldsymbol{R}_i$ and $\boldsymbol{p}_i$ can change continuously. However, in any real system, one must use the Heisenberg uncertainty principle, which effectively transforms this infinite number into a finite number of microstates.

Once we have defined the total number of microstates of an isolated system, we use the Boltzmann definition of entropy, in Eq. (2.7), to define the *fundamental equation* of an isolated system, by:

$$S(E, V, N) = k \ln \Omega(\mathrm{E}, V, N) \tag{2.7a}$$

This is a fundamental equation in the sense that all other thermodynamic quantities may be derived from it. Recall that $\Omega(\mathrm{E}, V, N)$ is the total number of micro-states of the system characterized by the variables (E, V, N). It follows from Eq. (2.7a) that also S (E, V, N) is the entropy of the (E, V, N) system at equilibrium.

The total differential of the function $S(E, V, N)$, is well known in thermodynamics. It has the form:

$$\begin{aligned} dS &= \left(\frac{\partial S}{\partial E}\right)_{V,N} dE + \left(\frac{\partial S}{\partial V}\right)_{E,N} dV + \left(\frac{\partial S}{\partial N}\right)_{E,V} dN \\ &= \frac{1}{T} dE + \frac{P}{T} dV - \frac{\mu}{T} dN \end{aligned} \tag{2.8}$$

In Eq. (2.8), T and P are the temperature and pressure, respectively, and μ is the chemical potential (if there are c-components, then μ is the vector $\boldsymbol{\mu} = (\mu_1, \mu_2, \ldots, \mu_c)$). Thus, having the fundamental equation, we can calculate T, P and μ of the system. One can continue and take second derivative of the fundamental equation to get other thermodynamic quantities of interest.

Following the second postulate of statistical thermodynamics we assert that if there are Ω micro-states, then the distribution $(Pr(1), \ldots, Pr(\Omega))$ must satisfy the normalization condition:

$$1 = \sum_{i=1}^{\Omega} Pr(i) = \sum_{i=1}^{\Omega} p = p\Omega \tag{2.9}$$

From (2.9) it follows that: $Pr(i) = p = \frac{1}{\Omega}$. Thus, in an ensemble of isolated systems, Fig. 2.4, the macro-state of each system is the same, (E, V, N). The micro-state could be different. The fraction of systems in the ensemble being in a microstate i, is $1/\Omega$.

2.4.2 *The Isothermal-Isochoric System; The* T, V, N *Ensemble*

In Sect. 2.4.1, we described an ensemble of M isolated systems, each of which is characterized by the variables (E, V, N), Fig. 2.4. We assumed that the systems in the ensemble do not interact with each other.

In most experiments, we do not use an isolated system but a system at a constant temperature. We now construct the so-called isothermal isochoric ensemble, referred to as the (T, V, N) *ensemble*, Fig. 2.5. To do so, we replace the thermally insulated boundaries (or athermal walls) in Fig. 2.4 by thermally conducting (or *diathermal*) boundaries. The ensemble as a whole, in Fig. 2.5, is still considered to be isolated. However, heat can *flow* between the individual systems in this ensemble. The volume V and the number of particles in each system is still fixed. As a result of the flow of heat between the systems, the value of the energy E will fluctuate, it will no longer be the same as in the ensemble in Fig. 2.4. We know from the Zeroth Law of thermodynamics, that the temperature must be constant throughout the entire ensemble.

The fundamental probability distribution in this ensemble is:

$$Pr(E) = \Omega(E, V, N)\exp[-\beta E]/Q \tag{2.10}$$

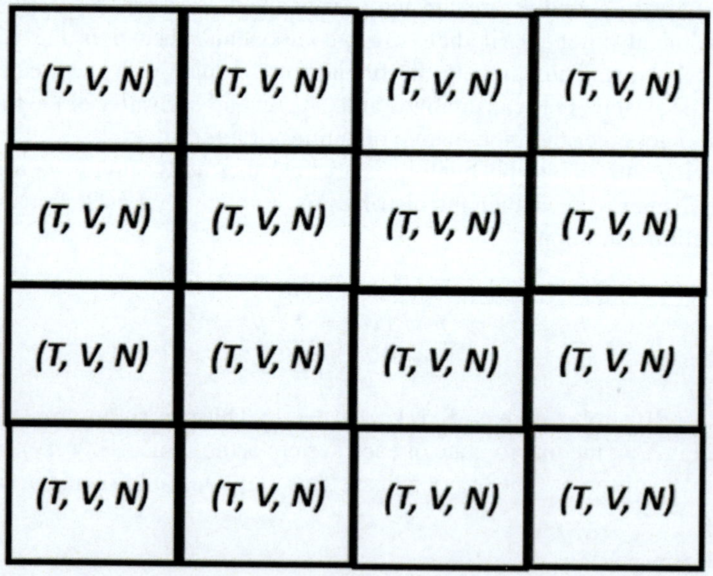

Fig. 2.5 An ensemble of isothermal isobaric systems. Each system is characterized by the same set of variables T, V and N

where β is used as a short-hand notation for $(kT)^{-1}$, and $Pr(E)$ is the probability of finding a system in the (*T, V, N*) ensemble having energy *E*. $\Omega(E, V, N)$, as before, is the number of microstates of a system having energy *E*, volume *V*, and number of particles *N*. *Q* is a normalization constant which is determined by the requirement that the sum over all possible energy "levels" must be one:

$$\sum_E Pr(E) = 1 \tag{2.11}$$

From (2.10) and (2.11) we obtain:

$$Q(T, V, N) = \sum_E \Omega(E, V, N)\exp[-\beta E] \tag{2.12}$$

The quantity $Q(T, V, N)$ is referred to as the *T, V, N canonical partition function* (PF). Note carefully that in (2.12) we started with the function $\Omega(E, V, N)$, i.e. the number of microstates of a system characterized by the variables *E, V, N*. By multiplying by the quantity $\exp[-\beta E]$ and summing over all values of *E*, the resulting quantity is no longer a function of the variable *E*. Instead, we introduced a new parameter β, or equivalently the temperature *T*. The resulting quantity is now a function of *T*, not of *E* (*V* and *N* remain unchanged).

The physical reason underlying this change in the thermodynamic variables is as follows. As we noted above, it follows from the Zeroth Law of thermodynamics that if we take two isolated systems and bring them to thermal equilibrium, heat will flow from the hot to the cold body. The energy of each system will no longer be fixed. The energy of each system will fluctuate, but at equilibrium, the two systems, will be characterized by a fixed temperature. In our case, we started with an ensemble of *M* systems, each characterized by the same energy *E* (as well as *V* and *N*). When we bring them to thermal contact, heat will be exchanged between the systems. The energy of each system will not be fixed, but the temperature of the entire ensemble will be constant. This is the (T, V, N) ensemble.

The fundamental thermodynamic function in this ensemble is the Helmholtz energy $A(T, V, N)$, and the fundamental equation is:

$$A(T, V, N) = -kT \ln Q(T, V, N) \tag{2.13}$$

Equation (2.13) is the *fundamental equation* of the (T, V, N) ensemble, or the *canonical ensemble*. Justification for this relationship may be found in any textbook on ST (e.g. Ben-Naim [1]). From Eq. (2.13) we can derive all the thermodynamic quantities of interest in the *T, V, N* ensemble. To do that, we used the well-known total differential of the Helmholtz energy (expressed as a function of the variables *T, V*, and *N*), which is:

$$dA = -SdT - PdV + \mu dN \tag{2.14}$$

From Eq. (2.14) we can obtain all the thermodynamic quantities:

$$S = \frac{-\partial A}{\partial T} = k \ln Q + kT \frac{\partial \ln Q}{\partial T} \tag{2.15}$$

$$P = \frac{-\partial A}{\partial V} = kT \frac{\partial \ln Q}{\partial V} \tag{2.16}$$

$$\mu = \frac{\partial A}{\partial N} = -kT \frac{\partial \ln Q}{\partial N} \tag{2.17}$$

2.4.3 *The Isothermal-Isobaric System; The* T, P, N *Ensemble*

The next change in the thermodynamic variable is from the volume V to the pressure P. In this step we "open" the boundaries of the systems so that *volume* can "flow" from one system to another, i.e., we replace the rigid boundaries by flexible or movable boundaries. The total volume of the entire ensemble is again fixed, but the volume of each system in the ensemble will fluctuate. Therefore, there will be a distribution of volumes. This distribution is continuous, but for simplicity, we assume that the volume of the system can attain only discrete values say $V_0, V_1, V_2, \ldots$. In this new ensemble the probability is of finding a single system having a specific volume V is:

$$Pr(V) = \frac{Q(T, V, N)\exp[-\beta PV]}{\Delta} \tag{2.18}$$

In Eq. (2.18) Δ is a new normalization constant, which is determined by the normalization condition:

$$\sum_V Pr(V) = 1 \tag{2.19}$$

From (2.18) and (2.19) we obtain the normalization constant:

$$\Delta(T, P, N) = \sum_V Q(T, V, N)\exp[-\beta PV] \tag{2.20}$$

Here, the summation is over all possible discrete values of the volume. If the volume changes continuously, then the sum in (2.20) is replaced by an integral, and the integral is the Laplace transform of $Q(T, V, N)$ with respect to the variable V. The quantity $\Delta(T, P, N)$ is referred to as the T, P, N *or the isothermal isobaric partition function.*

The fundamental thermodynamic function of this ensemble is the Gibbs energy, $G(T, P, N)$, and the fundamental relationship between the new partition function $\Delta(T, V, N)$ and thermodynamics is:

$$G(T, P, N) = -kT \ln \Delta(T, P, N) \tag{2.21}$$

where G is the Gibbs energy, expressed as a function of the variables T, P, N. From the fundamental relationship (2.21) we can obtain all the thermodynamic quantities of the system. To do this we write the well-known expression for the total differential of G, viewed as a function of T, P, N, which is:

$$dG = -SdT + VdP + \mu dN \tag{2.22}$$

From (2.22) we obtain all the other thermodynamic quantities of interest, e.g.:

$$S = -\frac{\partial G}{\partial T} = k \ln \Delta + kT \frac{\partial \ln \Delta}{\partial T} \tag{2.23}$$

$$H = G + TS = kT^2 \frac{\partial \ln \Delta}{\partial T} \tag{2.24}$$

In the theory of ST one usually proceed to the *open* system characterized by variables T, V, μ, where μ is the chemical potential (assuming there is only one component in the system). We shall not need this ensemble in this book. For details see Ben-Naim [1].

2.5 The "Classical" of Statistical Thermodynamics

In the previous sections of this chapter, we introduced the second postulate of statistical thermodynamics, and the Boltzmann definition of entropy using the language of quantum mechanics. However, in many applications of statistical thermodynamics, it is more convenient to describe a system *classically*, i.e. the *microstate* of a system is fully characterized by giving the locations and the momenta of all the atoms, or molecules of the system. With this characterization, we introduce the classical canonical partition function $Q(T, V, N)$ by analogy with the quantum mechanical case. The energy levels are replaced by the classical Hamiltonian function, i.e.

$$Energy\colon \boldsymbol{H}\left(\boldsymbol{p}^N, \boldsymbol{R}^N\right) = \sum_{i=1}^{N} \frac{\boldsymbol{p}_i^2}{2m} + U_N\left(\boldsymbol{R}^N\right) \tag{2.25}$$

Equation (2.25) means that for each set of parameters $\boldsymbol{p}_1, \boldsymbol{p}_2, \ldots, \boldsymbol{p}_N$ and $\boldsymbol{R}_1, \boldsymbol{R}_2, \ldots, \boldsymbol{R}_N$ there corresponds an energy "level" which consists of the kinetic energy of all the particles and the potential energy of interaction between all the particles. The "classical" analog of the canonical PF is thus:

$$Q(T, V, N) \approx \int \cdots \int \exp[-\beta \boldsymbol{H}(\boldsymbol{p}^N, \boldsymbol{R}^N)] d\boldsymbol{p}^N d\boldsymbol{R}^N \tag{2.26}$$

Here, integrations are carried out over all possible positions and momenta of the particles in the system. This quantity may be referred to as the "pure classical" PF. It turns out that this PF *does not* lead to results that conform with the experimental results. See Ben-Naim [1] for the so-called Gibbs paradox. Therefore, to achieve such an agreement, we must add two factors to the "pure classical" PF, the origin of which is in quantum mechanics; one is the Heisenberg *uncertainty principle*, and the second, due to the *indistinguishability* of the particles. Once we add these two corrections, we convert the "pure classical" PF, into what may be called the *classical limit* of the quantum mechanical PF, or simply the *classical* PF, which is:

$$Q(T, V, N) = \frac{q^N}{(8\pi^2)^N \Lambda^{3N} N!} \int \exp[-\beta U_N(\boldsymbol{X}^N)] d\boldsymbol{X}^N \tag{2.27}$$

where q is the internal partition function of a single molecule, Λ^{3N} is a result of integration over all possible momenta of the particles, and is given by:

$$\Lambda^3 = \left(\frac{h}{\sqrt{2\pi mkT}}\right)^3 \tag{2.28}$$

In (2.27), the vector, $\boldsymbol{X}^N = \boldsymbol{X}_1, \ldots, \boldsymbol{X}_N$ is the *configuration* of all the N particles, with $\boldsymbol{X}_i$ being the vector comprising the location and orientation of the i-th particle, and $d\boldsymbol{X}^N = d\boldsymbol{X}_1 d\boldsymbol{X}_2 \ldots d\boldsymbol{X}_N$.

With this PF, one does not get into the so called Gibbs paradox.

2.6 Some Results for an Ideal Gas

We now quote a few results obtained for the simplest system for which ST may be applied. This is an *ideal gas*. The equation of state of an ideal gas is well known:

$$PV = nRT \tag{2.29}$$

where n is the number of moles and R is referred to as the *gas constant*. A theoretical *ideal gas* is defined by the requirement that there are no intermolecular interactions among the particles, i.e.

$$U(\boldsymbol{X}^N) \equiv 0 \tag{2.30}$$

This means that for any configuration $\boldsymbol{X}^N \equiv (\boldsymbol{X}_1, \boldsymbol{X}_2, \ldots, \boldsymbol{X}_N)$, where $\boldsymbol{X}_i$ is the vector comprising the location ($\boldsymbol{R}_i$) and the orientation ($\boldsymbol{\Omega}_i$) of the particle i, the total interaction energy is zero. A theoretical ideal gas is only an idealized model for a real system with very weak intermolecular interactions. In real gases there are always interactions between the particles. However, when the density of the particles is extremely low, we can neglect all the intermolecular interaction energy in the system. In this limit we have a *real* or a *practical* ideal-gas system.

From the classical canonical partition function (PF), derived in Sect. 2.5 we have for an ideal gas:

$$\begin{aligned} Q(T, V, N) &= \frac{q^N}{(8\pi^2)^N \Lambda^{3N} N!} \int \cdots \int d\mathbf{X}^N \\ &= \frac{q^N}{(8\pi^2)^N \Lambda^{3N} N!} \left[\int_V d\mathbf{R} \int_0^{2\pi} d\phi \int_0^{\pi} \sin\theta d\theta \int_0^{2\pi} d\psi \right]^N \\ &= \frac{q^N V^N}{\Lambda^{3N} N!} \end{aligned} \quad (2.31)$$

Here, q is the internal PF of a single particle, and Λ^3 is the momentum PF. For simple spherical particles, sometimes referred to as "structure-less" particles, Eq. (2.31) reduces to:

$$Q(T, V, N) = V^N / \Lambda^{3N} N! \quad (2.32)$$

From the PF in (2.31) and the fundamental relationship (2.13), we can obtain all the thermodynamic quantities for an ideal gas. For instance, the expression for the pressure is:

$$P = kT \left(\frac{\partial \ln Q}{\partial V} \right)_{T,N} = \frac{kTN}{V} = \rho k T \quad (2.33)$$

where $\rho = N/V$ is the number density. This is the same as the equation of state (2.29). Here we used the Boltzmann constant k instead of the gas constant R, where $R = kN_{av}$, and N_{av} is the Avogadro number.

Another important quantity is the chemical potential of an ideal gas, which is:

$$\begin{aligned} \mu &= -kT \left(\frac{\partial \ln Q}{\partial N} \right)_{T,V} = kT \ln\left(\Lambda^3 q^{-1}\right) + kT \ln \rho \\ &= \mu^{0g}(T) + kT \ln \rho \end{aligned} \quad (2.34)$$

where $\rho = N/V$ is the number density and $\mu^{0g}(T)$ is referred to as the standard chemical potential. The entropy of an ideal gas may be obtained by differentiation of the PF with respect to T:

$$S = k \ln Q + kT\left(\frac{\partial \ln Q}{\partial T}\right)_{V,N}$$
$$= \frac{5}{2}kN - Nk \ln\left(\rho \Lambda^3 q^{-1}\right) + kTN\frac{\partial \ln q}{\partial T} \tag{2.35}$$

For simple particles ($q = 1$) this equation reduces to:

$$S = \frac{5}{2}kN - Nk \ln \rho \Lambda^3 \tag{2.36}$$

The internal energy of a system of simple particles is obtained from the thermodynamic equation:

$$E = A + TS$$
$$= kTN \ln \rho \Lambda^3 - kTN + T\left(\frac{5}{2}kN - Nk \ln \rho \Lambda^3\right)$$
$$= \frac{3}{2}kTN \tag{2.37}$$

In this case we see that the energy is entirely due to the kinetic energy of the particles. And $3kT/2$ is the average kinetic energy of one particle.

2.7 Molecular Distribution Functions

We now turn from an ideal gas to a liquid phase. We shall briefly introduce the concept of molecular distribution function (MDF). The MDF is basically a probability distribution function. We shall discuss mainly two MDF; the *singlet* and the *pair distribution functions*. The generalization to any number of particles is quite straightforward. We shall also discuss here mainly MDF in a canonical ensemble. Again, the extension to other ensembles is quite straightforward.

All the MDFs are derived from the fundamental probability distribution, in the canonical ensemble, which is:

$$Pr(\boldsymbol{X}^N) = \frac{\exp[-\beta U_N(\boldsymbol{X}^N)]}{\int \cdots \int \exp[-\beta U_N(\boldsymbol{X}^N)]d\boldsymbol{X}^N} \tag{2.38}$$

$Pr(\boldsymbol{X}^N)$ is the probability density of finding the N particles at a specific *configuration* denoted by: $\boldsymbol{X}^N = \boldsymbol{X}_1, \ldots, \boldsymbol{X}_N$. The configuration of each particle is described by the six-dimensional vector $\boldsymbol{X}$ consisting of three locational coordinates $\boldsymbol{R} = (x, y, z)$ (of some chosen point in the molecule), and three orientational coordinates $\Omega =$

(ϕ, θ, ψ) for the orientation of the entire molecule. To obtain the molecular distribution for n particles we have to integrate the distribution over all the configurations of all the other particles. Thus, all the MDF are in fact marginal probabilities.

2.7.1 *The Singlet Distribution Function in the Canonical Ensemble*

The simplest MDF is the *singlet* distribution function. We assume that the particles are rigid molecules, i.e. molecules without internal rotational degrees of freedom. We start with the distribution function $Pr(\boldsymbol{X}^N)$, in (2.29). If we integrate over the configurations of all $N - 1$ particles; 2, 3, ..., N we get the *specific* singlet MDF, i.e.

$$P^{(1)}(\boldsymbol{X}_1) = \int \cdots \int P(\boldsymbol{X}^N) d\boldsymbol{X}_2 \ldots d\boldsymbol{X}_N \tag{2.39}$$

Clearly, $P^{(1)}(\boldsymbol{X}_1)d\boldsymbol{X}_1$ is a marginal probability, this means the probability of finding the *specific* particle "1" in the element of configuration $d\boldsymbol{X}_1$ at $\boldsymbol{X}_1$.

Of course, we are never interested in the probability distribution of a *specific* particle. Therefore, we define the *generic* singlet MDF. Since the probability of finding particle i in $d\boldsymbol{X}'$ is independent of the specific particle denoted i, and since all the particles are identical the probability of finding *any* particle in the element of configuration $d\boldsymbol{X}'$ is simply N times the specific singlet distribution $P^{(1)}(\boldsymbol{X}')$. Hence, we define:

$$\begin{aligned} \rho^{(1)}(\boldsymbol{X}') &= N P^{(1)}(\boldsymbol{X}') \\ &= N \int \cdots \int Pr(\boldsymbol{X}', \boldsymbol{X}_2, \ldots, \boldsymbol{X}_N) d\boldsymbol{X}_2 \cdots d\boldsymbol{X}_N \end{aligned} \tag{2.40}$$

$\rho^{(1)}(\boldsymbol{X}')d\boldsymbol{X}'$ is the probability of finding *any* particle in the element $d\boldsymbol{X}'$. Note that the probability interpretation of $\rho^{(1)}(\boldsymbol{X}')d\boldsymbol{X}'$ is valid only when the element of volume (included in $d\boldsymbol{X}'$) is small enough so that no more than one particle can occupy this volume at any given time. A more general interpretation of $\rho^{(1)}(\boldsymbol{X}')$ as the local density of particles at the configuration $\boldsymbol{X}'$ is given below. For more details, see Ben-Naim [1, 2]. We start by first defining the average number of particles in the element of volume $d\boldsymbol{R}'$ by:

$$N(d\boldsymbol{R}') = d\boldsymbol{R}' \int \cdots \int Pr(\boldsymbol{X}^N) \sum_{i=1}^{N} \delta(\boldsymbol{R}_i - \boldsymbol{R}') d\boldsymbol{X}^N \tag{2.41}$$

From this it follows that the average local (number) density of particles in the element of volume $d\boldsymbol{R}'$ at $\boldsymbol{R}'$ is given by:

$$\rho^{(1)}(\boldsymbol{R}') = N(d\boldsymbol{R}')/d\boldsymbol{R}'$$
$$= \int \cdots \int Pr(\boldsymbol{X}^N) \sum_{i=1}^{N} (\boldsymbol{R}_i - \boldsymbol{R}') d\boldsymbol{X}^N \tag{2.42}$$

Note that $d\boldsymbol{R}'$ is an element of *volume* $dx'dy'dz'$, at $\boldsymbol{R}'$. The quantity $\rho^{(1)}(\boldsymbol{R}')$ is referred to as the *singlet molecular distribution function*. The meaning of $\rho^{(1)}(\boldsymbol{R}')$ as *local density* will prevail in all our applications. The normalization condition for the singlet MDF is:

$$\int_V \rho^{(1)}(\boldsymbol{R}') d\boldsymbol{R}' = N \int_V P^{(1)}(\boldsymbol{R}') d\boldsymbol{R}' = N \tag{2.43}$$

The last equality follows from the normalization of the probability distribution $P^{(1)}(\boldsymbol{R}')$, i.e. the probability of finding the specific particle 1 in *any* place in V is unity.

In a homogeneous fluid, we expect that $\rho^{(1)}(\boldsymbol{R}')$ will have the same value at any point $\boldsymbol{R}'$ within the system. (This is true apart from a very small region near the surface of the system which we always neglect in considering macroscopic systems). Therefore, we write:

$$\rho^{(1)}(\boldsymbol{R}') = \text{const.} \tag{2.44}$$

From (2.43) and (2.44) we obtain:

$$\text{const} \times \int_V d\boldsymbol{R}' = N \tag{2.45}$$

Therefore, we have the result:

$$\rho^{(1)}(\boldsymbol{R}') = N/V = \rho \tag{2.46}$$

The last equation is intuitively clear, it states that the local density at any point $\boldsymbol{R}'$ is equal to the bulk density ρ. In a similar way, we can define the singlet distribution function for location and orientation by:

$$\rho^{(1)}(\boldsymbol{X}') = \int \cdots \int Pr(\boldsymbol{X}^N) \sum_{i=1}^{N} \delta(\boldsymbol{X}_i - \boldsymbol{X}') d\boldsymbol{X}^N$$
$$= N \int \cdots \int Pr(\boldsymbol{X}^N) \delta(\boldsymbol{X}_1 - \boldsymbol{X}') d\boldsymbol{X}^N$$
$$= N P^{(1)}(\boldsymbol{X}') \tag{2.47}$$

In Eq. (2.47), $P^{(1)}(\mathbf{X}')$ is the probability density of finding a *specific* particle at a given configuration $\mathbf{X}'$. If the system is both homogeneous and isotropic fluid, then we must have:

$$\rho^{(1)}(\mathbf{X}') = \text{const.} \tag{2.48}$$

Using the normalization condition, that the total number of particles in the system is *N*, we obtain:

$$\int \rho^{(1)}(\mathbf{X}')d\mathbf{X}' = N \int P^{(1)}(\mathbf{X}')d\mathbf{X}' = N \tag{2.49}$$

Therefore, we have the final result for the singlet MDF:

$$\rho^{(1)}(\mathbf{X}') = N/(V8\pi^2) = \rho/8\pi^2 \tag{2.50}$$

The connection between $\rho^{(1)}(\mathbf{R}')$ and $\rho^{(1)}(\mathbf{X}')$ is obtained simply by integration over all the orientations of the particle:

$$\rho^{(1)}(\mathbf{R}') = \int \rho^{(1)}(\mathbf{X}')d\Omega' = \rho \tag{2.51}$$

2.7.2 *The Pair Distribution Function in the Canonical Ensemble*

As in the case of the singlet MDF, we must distinguish between the *specific* and the *generic* pair distribution functions. The former is easily obtained from the general distribution in Eq. (2.38). The second is more important in applications. As always we start with the fundamental probability density, $P(\mathbf{X}^N)$, defined in Eq. (2.38), in the T, V, N ensemble. For simple spherical particles, the *specific* pair distribution function is defined as the probability density of finding specific particle 1 at location $\mathbf{R}'$ and another specific particle 2 at $\mathbf{R}''$. For non-spherical particles, but having no internal rotations, the *specific* pair distribution function is defined as the probability density of finding particle 1 at $\mathbf{X}'$ and particle 2 at $\mathbf{X}''$. This can be obtained from $P(\mathbf{X}^N)$ by integrating over all the configurations of the remaining $N-2$ molecules. The result is:

$$P^{(2)}(\mathbf{X}', \mathbf{X}'') = \int \cdots \int Pr(\mathbf{X}', \mathbf{X}'', \mathbf{X}_3, \ldots, \mathbf{X}_N)d\mathbf{X}_3 \ldots d\mathbf{X}_N \tag{2.52}$$

$P^{(2)}(\mathbf{X}', \mathbf{X}'')d\mathbf{X}'d\mathbf{X}''$ is the probability of finding a *specific* particle, say 1 in the element $d\mathbf{X}'$ at $\mathbf{X}'$, and *another specific* particle, say 2, in the element $d\mathbf{X}''$ at $\mathbf{X}''$.

Since all particles are equivalent (in a one-component system), this probability is the same for any specific pair of two different particles.

To obtain the *generic pair* distribution function, we consider all the possible *specific* pair-events and their corresponding probabilities.

$$\begin{array}{ll} \textbf{Event} & \textbf{Probability} \\ 1 \text{ in } d\boldsymbol{X}' \text{ and } 2 \text{ in } d\boldsymbol{X}'' & P^{(2)}(\boldsymbol{X}', \boldsymbol{X}'')d\boldsymbol{X}'d\boldsymbol{X}'' \\ 1 \text{ in } d\boldsymbol{X}' \text{ and } 3 \text{ in } d\boldsymbol{X}'' & P^{(2)}(\boldsymbol{X}', \boldsymbol{X}'')d\boldsymbol{X}'d\boldsymbol{X}'' \\ \vdots & \\ 1 \text{ in } d\boldsymbol{X}' \text{ and } N \text{ in } d\boldsymbol{X}'' & P^{(2)}(\boldsymbol{X}', \boldsymbol{X}'')d\boldsymbol{X}'d\boldsymbol{X}'' \\ 2 \text{ in } d\boldsymbol{X}' \text{ and } 1 \text{ in } d\boldsymbol{X}'' & P^{(2)}(\boldsymbol{X}', \boldsymbol{X}'')d\boldsymbol{X}'d\boldsymbol{X}'' \\ \vdots & \end{array} \tag{2.53}$$

Since all particles are equivalent, the probabilities of all the $N(N-1)$ *specific* events listed on the left-hand-side of (2.53) are equal. The *generic* event is defined as:

$$\{\text{any particle in } d\boldsymbol{X}' \text{ and any other particle in } d\boldsymbol{X}''\} \tag{2.54}$$

The event in (2.54) is the union of all the $N(N-1)$ *specific* events listed on the left-hand side of Eq. (2.53). However, the probability of the event (2.54) is the sum of all probabilities of the events on the left-hand-side of (2.53) *only* if the latter events are disjoint. This condition will be met when the elements of volume $d\boldsymbol{R}'$ and $d\boldsymbol{R}''$ (each is part of the configuration $d\boldsymbol{X}'$ and $d\boldsymbol{X}''$, respectively), are small enough so that no more than one of the events in the list on the left-hand side of (2.53) may occur at any given time.

We are now ready to define the *generic* pair distribution function by:

$$\begin{aligned} &\rho^{(2)}(\boldsymbol{X}', \boldsymbol{X}'')d\boldsymbol{X}'d\boldsymbol{X}'' \\ &\quad = Pr\{\text{any particle in } d\boldsymbol{X}' \text{ and a different particle in } d\boldsymbol{X}''\} \\ &\quad = \sum_{i \neq j} Pr\{\text{particle } i \text{ in } d\boldsymbol{X}' \text{ and another particle } j \text{ in } d\boldsymbol{X}''\} \\ &\quad = \sum_{i \neq j} P^{(2)}(\boldsymbol{X}', \boldsymbol{X}'')d\boldsymbol{X}'d\boldsymbol{X}'' \\ &\quad = N(N-1)P^{(2)}(\boldsymbol{X}', \boldsymbol{X}'')d\boldsymbol{X}'d\boldsymbol{X}'' \end{aligned} \tag{2.55}$$

The last equality on the right-hand side of (2.55) follows from the equivalence of all the $N(N-1)$ pairs of specific and different particles. We now rewrite $\rho^{(2)}(\boldsymbol{X}', \boldsymbol{X}'')$ in such a way that it may be interpreted as an average quantity:

$$
\begin{aligned}
&\rho^{(2)}(\boldsymbol{X}',\boldsymbol{X}'')d\boldsymbol{X}'d\boldsymbol{X}''\\
&= N(N-1)d\boldsymbol{X}'d\boldsymbol{X}''\int\cdots\int Pr(\boldsymbol{X}',\boldsymbol{X}'',\boldsymbol{X}_3,\ldots,\boldsymbol{X}_N)d\boldsymbol{X}_3\ldots d\boldsymbol{X}_N\\
&= N(N-1)d\boldsymbol{X}'d\boldsymbol{X}''\int\cdots\int Pr(\boldsymbol{X}_1,\ldots,\boldsymbol{X}_N)\\
&\times\delta(\boldsymbol{X}_1-\boldsymbol{X}')\delta(\boldsymbol{X}_2-\boldsymbol{X}'')d\boldsymbol{X}_1\ldots d\boldsymbol{X}_N\\
&= d\boldsymbol{X}'d\boldsymbol{X}''\int\cdots\int Pr(\boldsymbol{X}^N)\sum_{\substack{i=1\\ i\neq j}}^{N}\sum_{j=1}^{N}\delta(\boldsymbol{X}_i-\boldsymbol{X}')\delta(\boldsymbol{X}_j-\boldsymbol{X}'')d\boldsymbol{X}^N
\end{aligned}
\tag{2.56}
$$

The various steps in Eqs. (2.56) are similar to the ones we did for the singlet MDF in Sect. 2.7.1. Note that we use the basic property of the Dirac delta function to extend the range of integration over all the vectors $\boldsymbol{X}_1,\ldots,\boldsymbol{X}_N$. We also used the equivalence of the N particles, to obtain an average of the quantity:

$$
d\boldsymbol{X}'d\boldsymbol{X}''\sum_{\substack{i=1\\ i\neq j}}^{N}\sum_{j=1}^{N}\delta(\boldsymbol{X}_i-\boldsymbol{X}')\delta(\boldsymbol{X}_j-\boldsymbol{X}'')
\tag{2.57}
$$

This quantity, when it is used in an integral, it *counts* the number of *pairs* of particles simultaneously occupying the two elements of configurations $d\boldsymbol{X}'$ and $d\boldsymbol{X}''$. It follows that, the final result in Eq. (2.56) is the *average number of pairs* occupying the two elements of configurations $d\boldsymbol{X}'$ and $d\boldsymbol{X}''$.

The normalization of $\rho^{(2)}(\boldsymbol{X}',\boldsymbol{X}'')$ follows directly from (2.56):

$$
\int\int\rho^{(2)}(\boldsymbol{X}',\boldsymbol{X}'')d\boldsymbol{X}'d\boldsymbol{X}'' = N(N-1)
\tag{2.58}
$$

Since we are in a closed system (fixed N), this is the *exact* number of pairs in the system. As in the previous section, we note that the meaning of $\rho^{(2)}(\boldsymbol{X}',\boldsymbol{X}'')$ as an average quantity is preserved upon integration over any region S. This is not the case, however, when its probabilistic meaning is adopted. For instance, the quantity:

$$
\int_S\int_S\rho^{(2)}(\boldsymbol{X}',\boldsymbol{X}'')d\boldsymbol{X}'d\boldsymbol{X}''
\tag{2.59}
$$

is, by definition, the *average number* of pairs occupying the region S (a factor of 1/2 should be included when we are interested in different pairs). This quantity cannot, in general, be interpreted as a probability. Another useful quantity, is the locational (or spatial) pair distribution function, defined by:

$$
\rho^{(2)}(\boldsymbol{R}',\boldsymbol{R}'') = \int\int\rho^{(2)}(\boldsymbol{X}',\boldsymbol{X}'')d\Omega'd\Omega''
\tag{2.60}
$$

In Eq. (2.60), the integration is carried out over the orientations of the two particles. Here, $\rho^{(2)}(\mathbf{R}', \mathbf{R}'')d\mathbf{R}'d\mathbf{R}''$ is the average number of pairs occupying $d\mathbf{R}'$ and $d\mathbf{R}''$ or, alternatively for infinitesimal elements $d\mathbf{R}'$ and $d\mathbf{R}''$, this is the probability of finding one particle in $d\mathbf{R}'$ at $\mathbf{R}'$, and a second particle in $d\mathbf{R}''$ at $\mathbf{R}''$.

Finally, we note that in this section we discussed the pair distributions for system of rigid particles having no internal rotational degrees of freedom. One can extend the definition of the MDF also for molecules having internal degrees of freedom, such as proteins. We shall not need these in the present book.

2.8 The Pair Correlation Function

Correlations are usually defined for two random variables, see Ben-Naim [5]. In this book we shall be interested in the statistical mechanical expression for the correlation between two *events* which are the locations of two particles. Suppose that $P(A)$ and $P(B)$ are the probabilities of the two events A and B respectively. We define the correlation between these two events by:

$$g(A, B) = \frac{\Pr(A \cap B)}{\Pr(A)\Pr(B)} \tag{2.61}$$

As usual, $g(A, B) = 1$, means that there is no correlation, when $g(A, B) > 1$, we say that there is *positive correlation,* and when $g(A, B) < 1$, we say there is *negative correlation between* A and B. It should be noted that the correlation defined here differs from the correlation defined in probability theory. In probability theory correlation is defined as the *difference* between the probability of the intersection and the product of the probabilities of each of the events. It is also normalized in such a way that its range of variation is between -1 to $+1$. For more details, see Ben-Naim [5].

Consider next the following two events: $\{\text{a particle in } d\mathbf{X}'\}$ and $\{\text{a particle in } d\mathbf{X}''\}$, and the *intersection* of these two events:

$$\{\text{a particle in } d\mathbf{X}'\} \; \textit{and} \; \{\text{a particle in } d\mathbf{X}''\} \tag{2.62}$$

We now seek a measure of the extent of dependence between these two events. In general, we expect that the two events in Eq. (2.62) are not independent. This means that the occurrence of one event may influence the probability of the occurrence of the second event.

In this book we are interested in "locational" events. If the separation $R = |\mathbf{R}' - \mathbf{R}''|$ between two particles is very large, we expect that the two events in (2.62) will be independent. In this case, we write the probability of their intersection of the two events as:

$$\begin{aligned}&Pr\{\{a\,particle\,in\,dX'\}\,and\,\{a\,particle\,in\,dX''\}\}\\&= Pr\{a\,particle\,in\,dX'\}\times Pr\{a\,particle\,in\,dX''\}\\&= \rho^{(1)}(X')dX'\rho^{(1)}(X'')dX'',\ \text{for } R\to\infty\end{aligned} \tag{2.63}$$

An equivalent way of writing Eq. (2.63) is:

$$\rho^{(2)}(X',X'')=\rho^{(1)}(X')\rho^{(1)}(X'')=\left(\rho/8\pi^2\right)^2,\ \text{for } R\to\infty \tag{2.64}$$

When Eq. (2.64) holds, we say that the local densities at X' and X'' are uncorrelated. For any finite distance, R, the probability density $\rho^{(2)}(X',X'')$ is, in general, not a product of the two singlet probabilities. This leads us to introduce a measure of dependence between the two events, which we denote by $g(X',X'')$:

$$\begin{aligned}\rho^{(2)}(X',X'')&=\rho^{(1)}(X')\rho^{(1)}(X'')g(X',X'')\\&=\left(\rho/8\pi^2\right)^2 g(X',X'')\end{aligned} \tag{2.65}$$

The first equality, on the right-hand side of (2.65) defines the correlation function. The second equality is valid for a homogeneous and isotropic fluid, see Sect. 2.7.1. We also define the *locational,* or the *spatial* pair correlation function by:

$$\rho^{(2)}(\boldsymbol{R}',\boldsymbol{R}'')=\rho^{(2)}g(\boldsymbol{R}',\boldsymbol{R}'') \tag{2.66}$$

The two correlations $g(\boldsymbol{R}',\boldsymbol{R}'')$ and $g(X',X'')$ are related by:

$$g(\boldsymbol{R}',\boldsymbol{R}'')=\frac{1}{\left(8\pi^2\right)^2}\int\int g(X',X'')d\Omega'\,d\Omega'' \tag{2.67}$$

We note that that the integral on the right-hand side of Eq. (2.67) has the formal structure of an "average" of $g(X',X'')$ over all the orientations of the two particles. This average is taken with the probability distribution, $d\Omega' d\Omega''/\left(8\pi^2\right)^2$. Clearly, at any finite distance, the probability of finding one particle in $d\Omega'$, and the second in $d\Omega''$, is not equal to $d\Omega' d\Omega''/\left(8\pi^2\right)^2$. It is only in the limit of very large distance, $|\boldsymbol{R}''-\boldsymbol{R}'|\to\infty$ that this probability becomes $d\Omega' d\Omega''/\left(8\pi^2\right)^2$.

One can easily generalize the concept of pair correlation for multi-component system. In such a case we need to distinguish between pair correlation for each pair of species α,β. For instance, if A and B are spherical particles, then we have three different pair correlation functions:

$$g_{AA}(R),\ g_{AB}(R)=g_{BA}(R),\ g_{BB}(R) \tag{2.68}$$

Note that $g_{AB}(R)=g_{BA}(R)$. An interesting and a useful interpretation of the pair correlation function is in terms of the *conditional probability* of observing a particle in dX'' at X'' given a particle at X', this is defined by:

$$\rho(X''|X')dX'' = \frac{\rho^{(2)}(X', X'')dX'dX''}{\rho^{(1)}(X')dX'}$$
$$= \rho^{(1)}(X'')g(X', X'')dX'' \tag{2.69}$$

The second equality on the right-hand side of Eq. (2.69) follows from the definition of $g(X', X'')$. Note that in the last expression on the right-hand side we dropped the element of configuration dX'. The reason is that the probability of finding a particle at an *exact* configuration X'' is zero, hence we must take an infinitesimal element of volume at X''. For the *conditional probability* we can specify the *exact condition*: "given a particle at X'."

Similarly, for the *locational pair* correlation function, we have the relationship:

$$\rho(R''|R') = \rho g(R', R'') \tag{2.70}$$

Which connects between the locational correlation function and the conditional density.

References

1. Ben-Naim A (1992) Statistical thermodynamics for chemists and biochemists. Plenum Press, New York
2. Ben-Naim A (2006) Molecular theory of solutions. Oxford University Press, Oxford, UK
3. Ben-Naim A (2017) Information theory, part I: an introduction to the fundamental concepts. World Scientific, Singapore
4. Ben-Naim A (2020) Entropy: the greatest blunder in the history of science. Independent Publisher, Amazon
5. Ben-Naim A (2015) Discover probability; how to use it, how to avoid misusing it and how it affects every aspect of your life. World Scientific, Singapore

Chapter 3
Solvent-Induced Interactions and Forces, Definitions and Connections with Thermodynamics

In this chapter, we introduce the two main concepts of this book; the solvent-induced interactions (SII) and the solvent-induced forces (SIF). The former is defined as part of the potential of mean force (PMF), and the second is defined as the derivative of the PMF. Following these definitions, we demonstrate the connection between the PMF and thermodynamic functions (either the Helmholtz energy or the Gibbs energy depending on whether we work is done in either *T, V, N*, or the *T, P, N* ensemble, respectively). Finally, we show how many confuse the two concepts of SII and SIF.

In most of this chapter, we will use the *T, V, N* ensemble. This is done mainly for the convenience of notation. Also, for convenience, we will assume that the solvent consists of *N* solid particles, each of which is fully described by the vector $\boldsymbol{X}$ consisting of three locational coordinates $\boldsymbol{R} = x, y, z$ and three orientation angles $\boldsymbol{\Omega} = (\phi, \theta, \psi)$. The solute will be assumed to be a simple spherical particle such as argon or hard sphere. In the case of pairwise interactions, we only need the locations of the two solute particles to describe their configuration. The corresponding PMF will be dependent only on the distance between the two particles. In more general cases where the solutes are rigid particles one should discuss the potential of mean force and torque instead of just the PMF.

In this chapter we will discuss the indirect interactions and forces in any solvent. However, in the rest of the book we shall be interested in aqueous solutions. Thus, whenever we refer to a solvent molecule we mean water molecules unless stated otherwise.

3.1 Definition of the Potential of Mean Force Between Two Simple Particles in a Solvent

Our system consists of *N* solvent molecules (usually water) contained in a volume *V* and at temperature *T*. A solvent molecule's configuration is described by the vector $\boldsymbol{X}$ consisting of six coordinates (3-locational and 3-orientational). The entire solvent's

A. Ben-Naim, *Solvent-Induced Interactions and Forces in Protein Folding*, Physical Chemistry in Action, https://doi.org/10.1007/978-3-031-31871-9_3

configuration will be denoted by $\boldsymbol{X}^N = \boldsymbol{X}_1, \ldots, \boldsymbol{X}_N$. In this system we have two simple spherical solute particles which we denote by A and B. These can either be different or identical particles. For simplicity we will assume that A and B are identical and their configuration is given by their locations $\boldsymbol{R}_A$ and $\boldsymbol{R}_B$. The whole system is assumed to be at constant T and V.

The solute–solute pair distribution function is given by:

$$Pr(\boldsymbol{R}_A, \boldsymbol{R}_B) = \frac{\int \exp[-\beta U(\boldsymbol{R}_A, \boldsymbol{R}_B, \boldsymbol{X}^N)]d\boldsymbol{X}^N}{\int \exp[-\beta U(\boldsymbol{R}_A, \boldsymbol{R}_B, \boldsymbol{X}^N)]d\boldsymbol{R}_A d\boldsymbol{R}_B d\boldsymbol{X}^N} \quad (3.1)$$

$Pr(\boldsymbol{R}_A, \boldsymbol{R}_B)$ is the probability density of finding the pair of particles A and B at the configuration $\boldsymbol{R}_A, \boldsymbol{R}_B$ (see Sect. 2.7). Note that the integration in the numerator of (3.1) is only over all the configurations of the solvent molecules. The integration in the denominator is over all the configurations of both solvent and solute molecules. Note also that since we have only two solute particles A and B we need not distinguish between the *specific* and the *generic* distribution function.

The pair correlation function is defined by the ratio of the two probabilities (or the probability densities):

$$g(\boldsymbol{R}_A, \boldsymbol{R}_B) = \frac{Pr(\boldsymbol{R}_A, \boldsymbol{R}_B)}{Pr(\boldsymbol{R}_A, \boldsymbol{R}_B, R_{AB} = \infty)} \quad (3.2)$$

where in the denominator on the right-hand side of Eq. (3.2), we have the pair probability density for two particles at infinite separation, $R_{AB} = \infty$. Clearly, at this distance we expect no correlation between the two locations of A and B, i.e. that $g(R_{AB} = \infty) = 1$.

We now use Eq. (3.1) to rewrite (3.2) as:

$$\begin{aligned} g(\boldsymbol{R}_A, \boldsymbol{R}_B) &= \frac{\int \exp[-\beta U(\boldsymbol{R}_A, \boldsymbol{R}_B, \boldsymbol{X}^N)]d\boldsymbol{X}^N}{\int \exp[-\beta U(R_{AB} = \infty, \boldsymbol{X}^N)]d\boldsymbol{X}^N} \\ &= \frac{Z_N(\boldsymbol{R}_A, \boldsymbol{R}_B)}{Z_N(R_{AB} = \infty)} \end{aligned} \quad (3.3)$$

In (3.3), $Z_N(\boldsymbol{R}_A, \boldsymbol{R}_B)$, and $Z_N(\boldsymbol{R}_{AB} = \infty)$ are the configurational partition function of N solvent molecules with two solute particles at the fixed configuration $(\boldsymbol{R}_A, \boldsymbol{R}_B)$, and at$(R_{AB} = \infty)$, respectively.

3.2 The Potential of Mean Force

The potential of mean force is *defined* in terms of the pair correlation function for the two solute particles, A and B, at $\boldsymbol{R}_A, \boldsymbol{R}_B$ by:

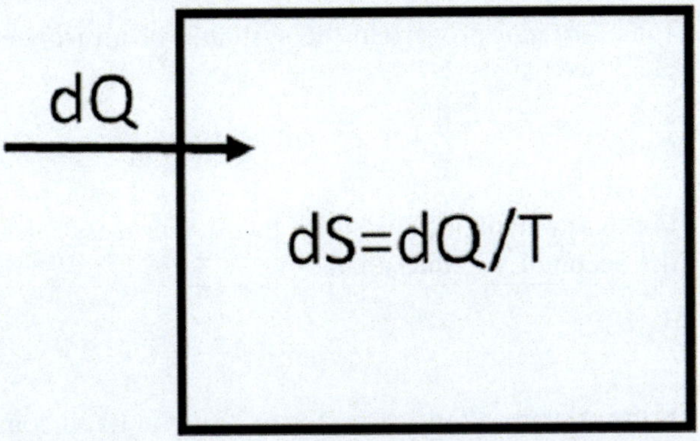

Fig. 3.1 Clausius's definition of a small change in entropy dS for the process of adding a small quantity of heat dQ to a system at a fixed temperature T

$$W(\boldsymbol{R}_A, \boldsymbol{R}_B) = -kT \ln g(\boldsymbol{R}_A, \boldsymbol{R}_B) \tag{3.4}$$

We now show that $W(\boldsymbol{R}_A, \boldsymbol{R}_B)$, as defined above may be interpreted as the work associated with the process of bringing A and B from fixed points at infinite separation to the final configuration $\boldsymbol{R}_A, \boldsymbol{R}_B$. Later we shall justify the term potential of mean *force*.

Before doing this, we digress to discuss some results from thermodynamics, relating *work* with either change in Helmholtz energy or change in Gibbs energy.

We recall Clausius' definition of the change in entropy associated with small heat transfer into a system at a temperature T:

$$dS = \frac{\delta Q}{T} \tag{3.5}$$

$\delta Q > 0$ means heat flows into the system, hence the entropy increases, Fig. 3.1.

When $\delta Q < 0$, heat flows out of the system and the entropy of the system decreases. Some people add the "condition" that the process is "reversible," and use the notation dS_{rev} in Eq. (3.5). In my view this additional condition is unnecessary. It is sufficient to say that δQ is a very small quantity of heat that is exchanged between the system and its surroundings, such that the temperature (T) of the system, is approximately constant in the process. If δQ is a large quantity of heat then the temperature of the system will be affected and the definition in Eq. (3.5) becomes invalidated.

In writing Eq. (3.5) it is also implicitly implied that the change in the entropy of the system is due entirely to the flow of heat. If there is any other spontaneous process in the system, then we write instead of Eq. (3.5):

$$dS = \frac{\delta Q}{T} + \delta S_{int} = d_e S + d_i S \tag{3.6}$$

where $d_e S$ is the *external* change of entropy, i.e., change due to the exchange of heat with the surroundings (or due to an *external* source), and $d_i S$ is the change due to some *internal* process in the system. For an isolated system (i.e., no exchange of heat), we have:

$$dS = d_i S \geq 0 \tag{3.7}$$

For thermodynamics we know that in a process carried out at constants T, V, and N, the Second Law states that:

$$\Delta A_{T,V,N} \leq 0 \tag{3.8}$$

If the system at T, V, N performs work on its surroundings, then:

$$\Delta A_{T,V,N} \geq W \tag{3.9}$$

Note that for any process between two equilibrium states of the system, the change ΔA, is fixed. However, the work done by the system depends, in general on the path of the process. It follows from Eq. (3.9), that the work done by the system is at most equal to $\Delta A_{T,V,N}$.

For a process at constant T, P, N, the analog of Eq. (3.9) is:

$$\Delta G_{T,P,N} \geq W_{net} = W - W_{\exp} \tag{3.10}$$

where W_{net} means the "net-work", i.e. all the work excluding expansion (PV) work.

This is the reason that either the Helmholtz or the Gibbs energy is sometimes referred to as the work function, work done either at constant T, V, N or T, P, N, respectively.

We now show that the PMF defined in Eq. (3.4) may be interpreted as a change in Helmholtz energy for the process of bringing the two solute particles from infinite separation to the final configuration $\mathbf{R}_A$, $\mathbf{R}_B$, while keeping T, V, and N constants.

In order to do this, we recall the relationship between the Helmholtz energy and the canonical PF (see Sect. 2.4.2). For our system of N solvent molecules and two solute particles at $\mathbf{R}_A$, $\mathbf{R}_B$ we can write the relationship between the PF and the Helmholtz energy of the system as:

$$A(\mathbf{R}_A, \mathbf{R}_B) = -kT \ln\left[\frac{q^N}{N!} \exp[-\beta U(\mathbf{R}_A, \mathbf{R}_B)] \times \int \cdots \int \exp[-\beta B(\mathbf{R}_A, \mathbf{R}_B) - \beta U_N(\mathbf{X}^N)] d\mathbf{X}^N\right] \tag{3.11}$$

Here, q is the *internal* the partition function of a single solvent molecule (including the momentum partition function) and we assume that the solute particles do not have any internal degrees of freedom. Additionally, since the two solute particles are at fixed position, they do not possess a momentum PF. The integration in Eq. (3.11) is over all the configurations of the solvent molecules. $U(\mathbf{R}_A, \mathbf{R}_B)$ is the pair potential

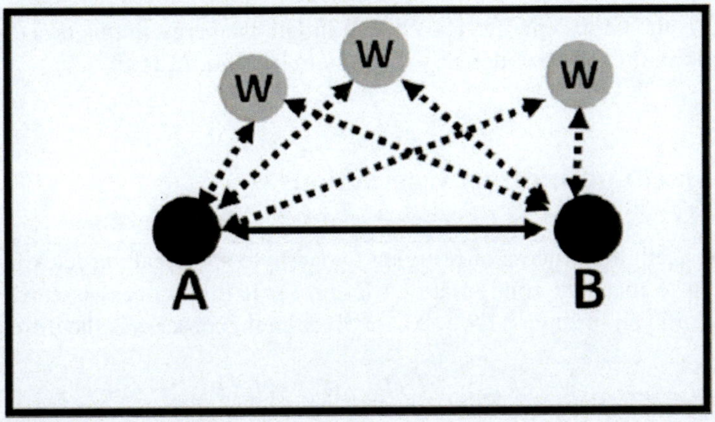

Fig. 3.2 The direct interaction energy between A and B is depicted by the solid arrow. The binding energy of A and B is depicted by the total interaction energies between A and B and all solvent (w) molecules shows as dashed arrows

for the pair of solute molecules at $(\boldsymbol{R}_A, \boldsymbol{R}_B)$ and $B(\boldsymbol{R}_A, \boldsymbol{R}_B)$ is the *binding energy* of the two solutes at $(\boldsymbol{R}_A, \boldsymbol{R}_B)$ to all the solvent molecules at configuration $\boldsymbol{X}^N$, Fig. 3.2.

Assuming pairwise additivity of the total potential energy, we write:

$$B(\boldsymbol{R}_A, \boldsymbol{R}_B) = \sum_{i=1}^{N} [U(\boldsymbol{R}_A, \boldsymbol{X}_i) + U(\boldsymbol{R}_B, \boldsymbol{X}_i)] \tag{3.12}$$

Thus, from (3.3), (3.4) and (3.11) we can rewrite PMF as:

$$W(\boldsymbol{R}_A, \boldsymbol{R}_B) = A(\boldsymbol{R}_A, \boldsymbol{R}_B) - A(R_{AB} = \infty) \tag{3.13}$$

where $A(\boldsymbol{R}_A, \boldsymbol{R}_B)$is the Helmholtz energy of the system of N solvent particles and two solute particles at a fixed configuration $\boldsymbol{R}_A, \boldsymbol{R}_B$. Therefore, $W(\boldsymbol{R}_A, \boldsymbol{R}_A)$ is the difference in the Helmholtz energy for the process (at constant T, V, N) of bringing the two solute particles from infinite separation, but at fixed positions to the final fixed-configuration $\boldsymbol{R}_A, \boldsymbol{R}_B$. Since ΔA for this process is associated with the maximum work done by the system we can also refer to $W(\boldsymbol{R}_A, \boldsymbol{R}_B)$ as the maximum work done by the system in this process. A negative value means that work is done by the system on the surroundings. A positive value means that the work is done on the system.

When the same process described above is carried out at constants T, P, N the analog of Eq. (3.13) is:

$$W(\boldsymbol{R}_A, \boldsymbol{R}_B) = G(\boldsymbol{R}_A, \boldsymbol{R}_B) - G(R_{AB} = \infty) \tag{3.14}$$

where G is the Gibbs energy. The change in Gibbs energy in this process is related to the maximum *net-work* done by the system [see Eq. (3.10)].

3.3 Direct and Indirect Interactions

Consider again the process of bringing two simple spherical particles from infinite separation to the final configuration $(\boldsymbol{R}_A, \boldsymbol{R}_B)$. If this process is carried out in a vacuum (or in an ideal gas) the work involved in this process is the difference:

$$W^{ig}(R_{AB}) = U(\boldsymbol{R}_A, \boldsymbol{R}_B) - U(R_{AB} = \infty) \tag{3.15}$$

We usually assume that the interaction between the two particles is zero at infinite separation. Hence, we identify the work function W with the pair interaction $U(R_{AB})$, where $R_{AB} = |\boldsymbol{R}_A - \boldsymbol{R}_B|$. If we do the same process in a liquid at T, V, N constants we can write the work associated with this process as:

$$\begin{aligned} W^{liq}(R_{AB}) &= A(\boldsymbol{R}_A, \boldsymbol{R}_B) - A(R_{AB} = \infty) \\ &= U(\boldsymbol{R}_A, \boldsymbol{R}_B) + \delta A(\boldsymbol{R}_A, \boldsymbol{R}_B) \end{aligned} \tag{3.16}$$

In Eq. (3.16), we divided the work into a sum of two parts; the *direct* interaction $U(\boldsymbol{R}_A, \boldsymbol{R}_B)$ (which we assume is the same as the interaction in a vacuum), and *indirect* interaction, or solvent-induced interaction SII.

From Eq. (3.11) we can write explicitly $\delta A(\boldsymbol{R}_A, \boldsymbol{R}_B)$ as:

$$\delta A(\boldsymbol{R}_A, \boldsymbol{R}_B) = -kT \ln \frac{\int \exp[-\beta B(\boldsymbol{R}_A, \boldsymbol{R}_B)] \exp[-\beta U_N] d\boldsymbol{X}^N}{\int \exp[-\beta B(R_{AB} = \infty)] \exp[-\beta U_N] d\boldsymbol{X}^N} \tag{3.17}$$

Thus, the total interaction between A and B is the sum of the direct interaction between A and B (see bold arrow in Fig. 3.3), and an indirect interaction mediated through the solvent (see dashed arrows in Fig. 3.3, for a "solvent" consisting of one water molecule W).

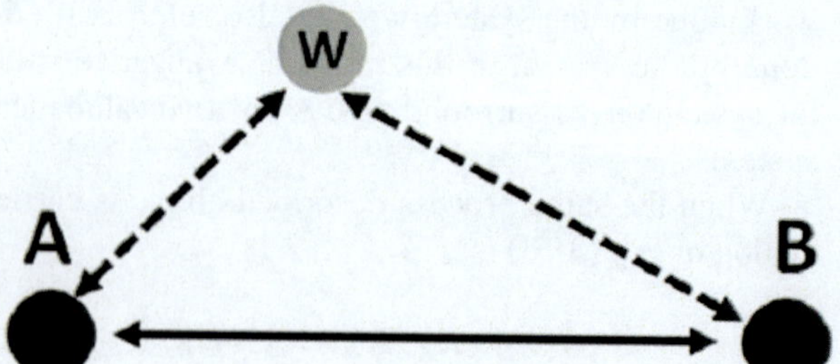

Fig. 3.3 Direct (solid arrow) and indirect interactions (dashed arrows) between A and B

There exists a well-known density expansion of the PMF (or of the pair correlation function) which takes into account interactions mediated by one, two, three, etc. solvent molecules. This is referred to as the cluster expansion. We shall always be concerned with the total indirect (or solvent induced) interactions, and not in individual contributions of various clusters of solvent molecules.

Another very useful interpretation of δA is in terms of the difference in solvation Helmholtz energy. To see this, we expand the ratio in Eq. (3.17) by multiplying and dividing it by configurational PF of the solvent:

$$Z_N = \int \cdots \int \exp[-\beta U_N] d\boldsymbol{X}^N \tag{3.18}$$

Remember that the probability distribution of configurations of the solvent molecules in the absence of solute molecules is:

$$Pr(\boldsymbol{X}^N) = \frac{\exp[-\beta U_N]}{Z_N} \tag{3.19}$$

We can now rewrite Eq. (3.17) as:

$$\begin{aligned} \delta A(\boldsymbol{R}_1, \boldsymbol{R}_2) = &-kT \ln\langle \exp[-\beta B(\boldsymbol{R}_A, \boldsymbol{R}_B)]\rangle_0 \\ &+ kT \ln\langle \exp[-\beta B(R_{AB} = 0)]\rangle_0 \end{aligned} \tag{3.20}$$

The two terms on the right-hand-side of (3.20) are the solvation Helmholtz energies of the pair of solute molecules at the initial and final configurations. Note that each solvation Helmholtz energy is an average of the quantity $\exp[-\beta B]$, the average is taken over all configurations of the solvent molecules of pure water (i.e. the absence of solute molecule). This average is indicated by the subscript zero in $\langle\ \rangle_0$. The meaning of δA as the difference in the solvation Helmholtz energies is evident by looking at Fig. 1.7

Note further that since the events "particle at A and $\boldsymbol{R}_A$" and "particle B at $\boldsymbol{R}_B$" become independent when $R_{AB} = \infty$, we can rewrite the second term on the right-hand-side of (3.20) as:

$$\begin{aligned} &kT \ln\langle \exp[-\beta B(\boldsymbol{R}_{AB} = \infty)]\rangle_0 \\ &\quad = kT \ln\{\langle \exp[-\beta B(\boldsymbol{R}_A)]\rangle_0 \langle \exp[-\beta B(\boldsymbol{R}_B)]\rangle_0\} \end{aligned} \tag{3.21}$$

Hence, we can write the final form for the indirect (solvent-induced interaction) as:

$$\delta A(\boldsymbol{R}_A, \boldsymbol{R}_B) = \Delta A^*(\boldsymbol{R}_A, \boldsymbol{R}_B) - \Delta A^*(\boldsymbol{R}_A) - \Delta A^*(\boldsymbol{R}_B) \tag{3.22}$$

where $\Delta A^*(\alpha)$ is the solvation Helmholtz energy of the "solute" α.

So far we discussed the process of bringing two solute particles from infinite separation to the distance R_{AB} in a solvent of constants T, V, N.

We could do the same process keeping T, P, N constants. In this case the analog of (3.22) would be:

$$\delta G(\boldsymbol{R}_A, \boldsymbol{R}_B) = \Delta G^*(\boldsymbol{R}_A, \boldsymbol{R}_B) - \Delta G^*(\boldsymbol{R}_A) - \Delta G^*(\boldsymbol{R}_B) \tag{3.23}$$

where $\Delta G^*(\alpha)$ is the solvation Gibbs energy of the solute α. For more details, see Ben-Naim [1].

One can easily generalize to any number of solute particles, and also for soluble particles which are rigid (i.e. the configuration of which requires both locational and angular coordinates).

3.4 Direct and Indirect Forces

In this section, we shall derive an important and very useful expression for the indirect (or solvent induced) force. This expression will justify the term "potential of mean *force*" (PMF) on one hand, and it will be essential for the discussion of the *speed* of the protein folding process on the other hand.

The force acting, on say, particle A by particle B when two particles are in vacuum is obtained by taking the gradient of the potential function $U(\boldsymbol{R}_A, \boldsymbol{R}_B)$ with respect to the vector $\boldsymbol{R}_A$, i.e.

$$\boldsymbol{F}_{A,direct} = -\nabla_A U(\boldsymbol{R}_A, \boldsymbol{R}_B) \tag{3.24}$$

In Eq. (3.24) ∇_A is the gradient with respect to the vector $\boldsymbol{R}_A = (x_A, y_A, z_A)$, defined by:

$$\nabla_A = \left(\frac{\partial}{\partial x_A}, \frac{\partial}{\partial y_A}, \frac{\partial}{\partial z_A}\right) \tag{3.25}$$

When the two particles are simple and spherical, the pair-potential becomes a function of the distance between the two particles, and the gradient reduces to a derivative with respect to the distance R_{AB}.

We now take the gradient of the PMF, Eq. (3.16), and we will get the total *force* acting on A:

$$\boldsymbol{F}_A = -\nabla_A W(\boldsymbol{R}_A, \boldsymbol{R}_B) = -\nabla_A U(\boldsymbol{R}_A, \boldsymbol{R}_B) - \nabla_A \delta A(\boldsymbol{R}_A, \boldsymbol{R}_B) \tag{3.26}$$

The first term on the right-hand-side of (3.26) is the direct force on A due to the presence of B at $\boldsymbol{R}_B$. The second term is the solvent-induced force.

We use Eq. (3.17) to obtain the indirect force:

$$\nabla_A \delta A(\boldsymbol{R}_A, \boldsymbol{R}_B) = \frac{\int \exp[-\beta B(\boldsymbol{R}_A, \boldsymbol{R}_B)] \exp[-\beta U_N] \nabla_A B(\boldsymbol{R}_A, \boldsymbol{R}_B) d\boldsymbol{X}^N}{\int \exp[-\beta B(\boldsymbol{R}_A, \boldsymbol{R}_B)] \exp[-\beta U_N] d\boldsymbol{X}^N} \tag{3.27}$$

Since the binding energy is given by:

$$B(\boldsymbol{R}_A, \boldsymbol{R}_B) = \sum_{i=1}^{N} [U(\boldsymbol{R}_A, \boldsymbol{X}_i) + U(\boldsymbol{R}_B, \boldsymbol{X}_i)] \tag{3.28}$$

And since the gradient ∇_A operates only on $\boldsymbol{R}_A$, we have:

$$\nabla_A B(\boldsymbol{R}_A, \boldsymbol{R}_B) = \sum_{i=1}^{N} \nabla_A U(\boldsymbol{R}_A, \boldsymbol{X}_i) \tag{3.29}$$

Note that the solvation Helmholtz energy of A in Eq. (3.22) is independent of the specific location $\boldsymbol{R}_A$. The notation $\Delta A^*(\boldsymbol{R}_A)$, is used to indicate that we have a particle A at a specific location, but $\Delta A^*(\boldsymbol{R}_A)$ is not a function of $\boldsymbol{R}_A$. Therefore, the gradient does not operate on $\Delta A^*(\boldsymbol{R}_A)$.

We next define the conditional probability distribution (or the conditional density) of solvent molecules at $\boldsymbol{X}^N$, given two solute particles at fixed positions $(\boldsymbol{R}_A, \boldsymbol{R}_B)$.

$$Pr\left(\boldsymbol{X}^N | \boldsymbol{R}_A, \boldsymbol{R}_B\right) = \frac{Pr\left(\boldsymbol{X}^N, \boldsymbol{R}_A, \boldsymbol{R}_B\right)}{Pr(\boldsymbol{R}_A, \boldsymbol{R}_B)} \tag{3.30}$$

where

$$Pr\left(\boldsymbol{X}^N, \boldsymbol{R}_A, \boldsymbol{R}_B\right) = \frac{\exp\left[-\beta U\left(\boldsymbol{X}^N, \boldsymbol{R}_A, \boldsymbol{R}_B\right)\right]}{\int \exp\left[-\beta U\left(\boldsymbol{X}^N, \boldsymbol{R}_A, \boldsymbol{R}_B\right)\right] d\boldsymbol{X}^N d\boldsymbol{R}_A d\boldsymbol{R}_B} \tag{3.31}$$

$$Pr(\boldsymbol{R}_A, \boldsymbol{R}_B) = \frac{\int \exp\left[-\beta U\left(\boldsymbol{X}^N, \boldsymbol{R}_A, \boldsymbol{R}_B\right)\right] d\boldsymbol{X}^N}{\int \exp\left[-\beta U\left(\boldsymbol{X}^N, \boldsymbol{R}_A, \boldsymbol{R}_B\right)\right] d\boldsymbol{X}^N d\boldsymbol{R}_A d\boldsymbol{R}_B} \tag{3.32}$$

Note that in the denominators of Eqs. (3.31) and (3.32) the integration is over all the $N + 2$ particles. However, when we form the conditional probability in (3.30) this denominator cancels out, and we get:

$$Pr\left(\boldsymbol{X}^N | \boldsymbol{R}_A, \boldsymbol{R}_B\right) = \frac{\exp\left[-\beta U\left(\boldsymbol{X}^N, \boldsymbol{R}_A, \boldsymbol{R}_B\right)\right]}{\int \exp\left[-\beta U\left(\boldsymbol{X}^N, \boldsymbol{R}_A, \boldsymbol{R}_B\right)\right] d\boldsymbol{X}^N} \tag{3.33}$$

In Eq. (3.33) the integration is over all the configurations of the solvent molecules. The *total* potential energy of the system is written as:

$$U\left(\boldsymbol{X}^N, \boldsymbol{R}_A, \boldsymbol{R}_B\right) = U(\boldsymbol{R}_A, \boldsymbol{R}_B) + B(\boldsymbol{R}_A, \boldsymbol{R}_B) + U_N\left(\boldsymbol{X}^N\right) \tag{3.34}$$

When substituting (3.34) in (3.33), the direct interaction $U(\boldsymbol{R}_A, \boldsymbol{R}_B)$ is canceled out and what remains is:

$$Pr\left(\boldsymbol{X}^N|\boldsymbol{R}_A, \boldsymbol{R}_B\right) = \frac{\exp[-\beta B(\boldsymbol{R}_A, \boldsymbol{R}_B)]\exp[-\beta U_N]}{\int \exp[-\beta B(\boldsymbol{R}_A, \boldsymbol{R}_B)]\exp[-\beta U_N]d\boldsymbol{X}^N} \tag{3.35}$$

It is now clear that the right-hand-side of Eq. (3.27) is an average of the quantity $\nabla_A B(\boldsymbol{R}_A, \boldsymbol{R}_B)$ with the *conditional probability distribution* given in (3.35), i.e. we write:

$$\begin{aligned} -\nabla_A \delta A(\boldsymbol{R}_A, \boldsymbol{R}_B) &= \int \{-\nabla_A B(\boldsymbol{R}_A, \boldsymbol{R}_B)\} Pr\left(\boldsymbol{X}^N|\boldsymbol{R}_A, \boldsymbol{R}_B\right) d\boldsymbol{X}^N \\ &= \langle -\nabla_A(\boldsymbol{R}_A, \boldsymbol{R}_B)\rangle \end{aligned} \tag{3.36}$$

We can further simplify the average quantity in Eq. (3.36) by noting that the gradient ∇_A operates only on $\boldsymbol{R}_A$, hence using Eq. (3.29) we can rewrite (3.36) as:

$$\begin{aligned} -\nabla_A \delta A(\boldsymbol{R}_A, \boldsymbol{R}_B) &= \int \left\{-\nabla_A \sum_{i=1}^{N} U(\boldsymbol{R}_A, \boldsymbol{X}_i)\right\} Pr\left(\boldsymbol{X}^N|\boldsymbol{R}_A, \boldsymbol{R}_B\right) d\boldsymbol{X}^N \\ &= \sum_{i=1}^{N} \int \{-\nabla_A U(\boldsymbol{R}_A, \boldsymbol{X}_i)\} Pr\left(\boldsymbol{X}^N|\boldsymbol{R}_A, \boldsymbol{R}_B\right) d\boldsymbol{X}^N \\ &= N \int \{-\nabla_A U(\boldsymbol{R}_A, \boldsymbol{X}_W)\} d\boldsymbol{X}_W \int Pr\left(\boldsymbol{X}^N|\boldsymbol{R}_A, \boldsymbol{R}_B\right) d\boldsymbol{X}^{N-1} \\ &= \int \{-\nabla_A U(\boldsymbol{R}_A, \boldsymbol{X}_W)\} N Pr(\boldsymbol{X}_W|\boldsymbol{R}_A, \boldsymbol{R}_B) d\boldsymbol{X}_W \\ &= \int [-\nabla_A U(\boldsymbol{R}_A, \boldsymbol{X}_W)] \times [\rho(\boldsymbol{X}_W|\boldsymbol{R}_A, \boldsymbol{R}_B)] d\boldsymbol{X}_W \end{aligned} \tag{3.37}$$

The reader is urged to check carefully each step in the procedure of obtaining the right-hand side of Eq. (3.37). We started by substituting Eq. (3.29) into (3.36). Next, we took the sum over i outside from the integral. Since all of the water molecules are equivalent, we can choose one specific water molecule, say, number 1 and replace the sum over the index i by N times the same integral. We next denoted the configuration of the water molecules number 1 by $\boldsymbol{X}_W$. The conditional probability of finding water molecule number 1 at $\boldsymbol{X}_W$, given that the two solutes are at $(\boldsymbol{R}_A, \boldsymbol{R}_B)$ is denoted by $Pr(\boldsymbol{X}_W|\boldsymbol{R}_A, \boldsymbol{R}_B)$. Finally, we multiply this (*specific*) conditional probability by N to obtain the *generic* probability density, i.e. $\rho(\boldsymbol{X}_W|\boldsymbol{R}_A, \boldsymbol{R}_B)d\boldsymbol{X}_W$ is the probability of finding *any* water molecule within $d\boldsymbol{X}_W$ at $\boldsymbol{X}_W$, given the two solutes at the fixed configuration $(\boldsymbol{R}_A, \boldsymbol{R}_B)$. This is also the *conditional local density* of water molecules at $\boldsymbol{X}_W$, given the two solutes particles at $(\boldsymbol{R}_A, \boldsymbol{R}_B)$.

The last form on the right-hand side of Eq. (3.37) is most useful in interpreting the indirect, or the solvent-induced force on $\boldsymbol{R}_A$ given the two solutes at $\boldsymbol{R}_A$, $\boldsymbol{R}_B$, and averaged over all possible solvent configurations. This equation was first derived in the context of simple liquids by Ben-Naim [1]. Later, it was applied to aqueous solutions. We rewrite the final result again:

$$-\nabla_A \delta A(\boldsymbol{R}_A, \boldsymbol{R}_B) = \int [-\nabla_A U(\boldsymbol{R}_A, \boldsymbol{X}_W)] \times [\rho(\boldsymbol{X}_W | \boldsymbol{R}_A, \boldsymbol{R}_B)] d\boldsymbol{X}_W \quad (3.38)$$

The integrand in (3.38) consists of two factors. The first, is the "direct" force exerted by a single water molecule at $\boldsymbol{X}_W$ on the solute particle at $\boldsymbol{R}_A$. The second term is the probability, or the conditional local density of water molecules to be found at $\boldsymbol{X}_W$, given the two solute particles at $(\boldsymbol{R}_A, \boldsymbol{R}_B)$. Clearly, in order to have a large solvent-induced force on A, we need some region in the configurational space of a water molecule for which both factor in this integrand will be large.

We can now rewrite the total force on A using Eqs. (3.26) and (3.38) as:

$$\begin{aligned} \boldsymbol{F}_A(total) &= -\nabla_A W(\boldsymbol{R}_A, \boldsymbol{R}_B) = \boldsymbol{F}_A(direct) + F_A(indirect) \\ &= -\nabla_A U(\boldsymbol{R}_A, \boldsymbol{R}_B) - \nabla_A \delta A(\boldsymbol{R}_A, \boldsymbol{R}_B) \\ &= -\nabla_A \delta U(\boldsymbol{R}_A, \boldsymbol{R}_B) + \int [-\nabla_A U(\boldsymbol{R}_A, \boldsymbol{X}_W)] \times [\rho(\boldsymbol{X}_W | \boldsymbol{R}_A, \boldsymbol{R}_B)] d\boldsymbol{X}_W \end{aligned} \quad (3.39)$$

It is helpful to think of the solvent-induced force by starting with an ideal gas solvent (i.e. $\rho_W \rightarrow 0$), and gradually increase the density of the solvent. Initially, when ρ_W is very small, no matter how strong the force $-\nabla_A U(\boldsymbol{R}_A, \boldsymbol{X}_W)$ is, the local density at any specific $\boldsymbol{X}_W$ is very small, and therefore we cannot expect any large solvent-induced force. As we increase the solvent density, we also increase the local conditional density at $\boldsymbol{X}_W$. In some actual examples discussed in in this book, we shall also examine the effect of the *condition*, i.e. the presence of A and B on the local density at $\boldsymbol{X}_W$. As we shall see in Chap. 6, Eq. (3.38) for the solvent-induced force is quite useful even for a qualitative argument regarding the question of dominant $H\phi I$ forces in protein folding, hence also answering the so-called Levinthal's Paradox. It should be noted that Eq. (3.39) is also valid in a system at constants *T*, *P*, *N*. In this ensemble, we should replace the Helmholtz energy in all the previous equation (Eq. 3.22) by the Gibbs energy, and also reinterpret the local conditional density $\rho(\boldsymbol{X}_W | \boldsymbol{R}_A, \boldsymbol{R}_B)$ as a density in the *T*, *P*, *N* ensemble.

A straightforward generalization of Eq. (3.38) should be mentioned. For solutes which are not spherical we should use instead of the gradient with respect to $\boldsymbol{R}_A$, the more general gradient with respect to the six-dimensional vector $\boldsymbol{X}_A$. In such a case, $W(\boldsymbol{X}_A, \boldsymbol{X}_B)$ will be referred to as the potential of mean force and torque. We shall not need the torque explicitly in our discussion. If we have a protein and we take the derivative with respect to the angle ϕ, we get a torque on the whole protein molecule. However, if we take the gradient with respect to $\boldsymbol{R}_i$, where i is a small group, say, a methyl group on the protein, we shall get a *force* on i, but that force will be a torque on the entire protein Fig. 3.4.

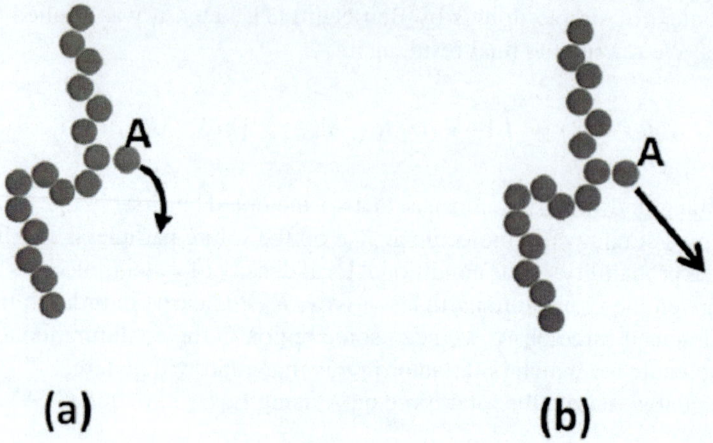

Fig. 3.4 **a** A gradient with respect to an angle will produce a torque on the entire protein. **a** A gradient with respect to the locational vector of group A on the protein will produce a force on that group

Another generalization of Eq. (3.38) is when we start with a group of n solute particles (or groups attached to a protein). In such a case, the condition $(\boldsymbol{R}_A, \boldsymbol{R}_B)$ will be replaced by the more general condition $(\boldsymbol{R}_1, \boldsymbol{R}_2, \ldots, \boldsymbol{R}_n)$. This generalization is very important when we are interested in the solvent induce force on a specific group of a protein, given a specific configuration (or conformation) of the entire protein. We shall discuss these generalizations in Sect. 3.4.

It should be emphasized that the literature is full of misconceptions and confusions about the "interactions," "driving forces" and forces. First, one should make a clear-cut distinction between *force* and *interaction*. When one discusses the stability of protein the factors which are important are *interactions*. On the other hand, when we discuss the *speed* of the folding process, then one should focus on the *forces* exerted on the various groups along the protein. In general, strong interactions do not imply strong forces, as is demonstrated below.

Because of its utmost importance, we provide a few examples to demonstrate the difference between the extent of interaction $U(r)$ and the force. We normally associate *strong force* with *strong interaction*. This is "normal" because we feel that if the slope of the curve $U(r)$ is larger or steeper, then we expect to get lower values of $U(r)$. This "normal" conclusion is shown in Fig. 3.5a, where we drew two Lennard–Jones potential functions; one with a deeper minimum than the other. Focusing on the attractive part of the curve (i.e. when the slope is positive) we see that the slope at each point on the lower curve is larger than that on the higher curve. The derivatives of the two curves in Fig. 3.5a are shown in Fig. 3.4b. It is clear that in this case strong interactions are associated with strong forces.

However, this is not always true. In Fig. 3.6 we show an example of two interaction curves. At a distance r_a the curve with a lower value of $U(r)$ has a smaller slope, i.e. weaker force.

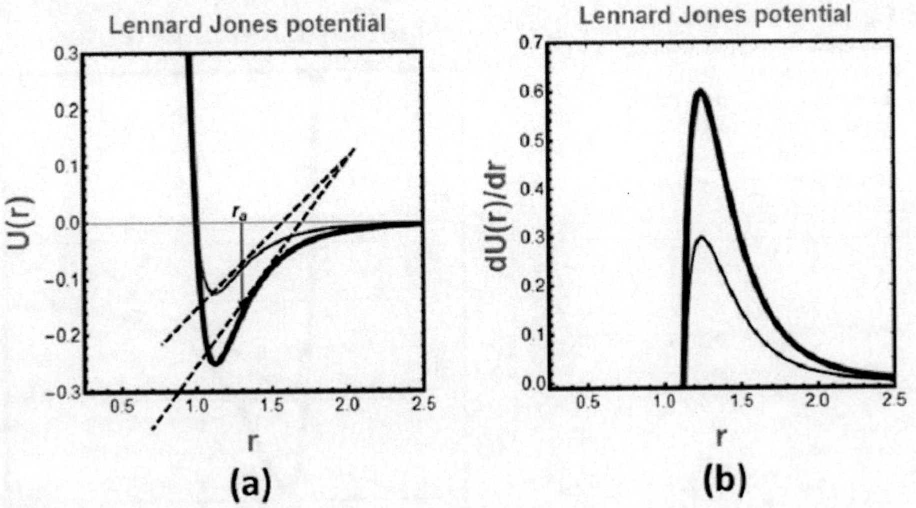

Fig. 3.5 **a** Two potential functions. At each distance *r*, the strength of the interaction is larger in one curve than the other. **b** The same is true for the force at each point which is the same as the derivative of *U(r)* with respect to the distance *r*

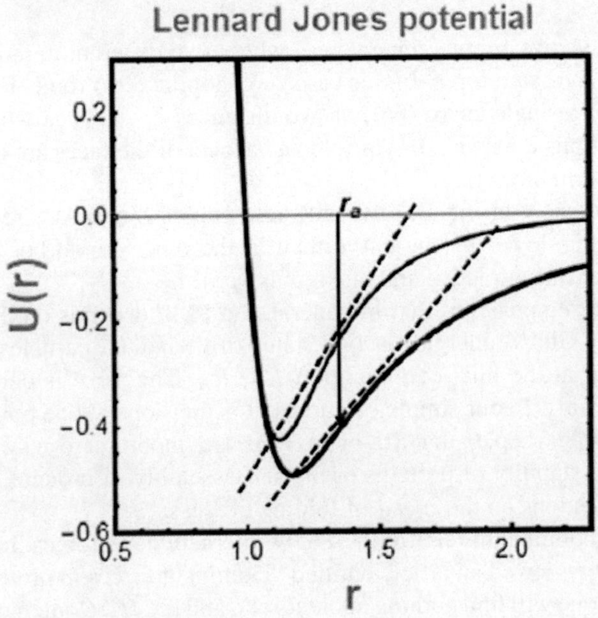

Fig. 3.6 Two potential functions *U(r)*. At the point r_a the interaction energy in the lower curve is larger than that in the upper curve. At the same distance, r_a the force in the lower curve is weaker than the force in the upper curve

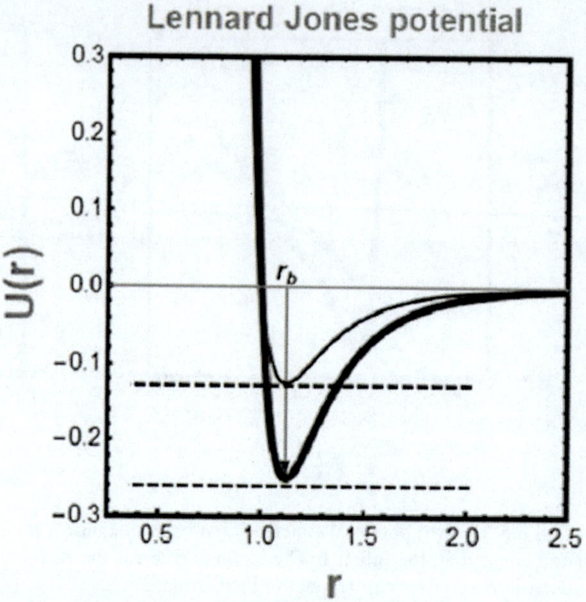

Fig. 3.7 Two potential functions $U(r)$. At the point r_b the interaction energy in the lower curve is larger than that in the upper curve. At the same distance the forces in both curves are equal to zero

Figure 3.7 shows a more general case where two different minima in the curves of $U(r)$ have the same force. In this case zero slope, i.e. no force. Figure 3.8 shows an example of a single curve $U(r)$ at two distances r_a and r_b at which the strength of the interactions are equal, but the slopes, hence, of the faces are different both in magnitude and in the sign.

Finally, we show in Fig. 3.9 two different curves $U(r)$. At a separation r_a, the value of $U(r)$ is lower in one curve than in the other curve, but the forces have different signs; one attractive and one repulsive.

It should be emphasized that in general, the PMF depends on the difference in Helmholtz (or Gibbs) energies at two states of the solute particles; i.e. at infinite separation and at the final configuration $\boldsymbol{R}_A, \boldsymbol{R}_B$. The *force* is determined at one configuration only. In our simple examples, it is the slope at one point.

As shall see in Chap. 6, that SII, or the PMF are important to answering questions relevant to the stability of proteins or the self-assembly of proteins. The forces are relevant to questions on the *speed* of folding process.

It is not uncommon to refer to the $H\phi O$ interactions as forces. In 1990, Dill has published and reviewed an article entitled "Dominant forces in protein folding," [ ]. In this review one will find nothing about *forces*, and the $H\phi O$ interactions discussed in that review are anything but "dominant" force (or interaction).

Sometimes, the Gibbs (or Helmholtz) energy difference [$\Delta A(\boldsymbol{R}_A, \boldsymbol{R}_B)$ or $\Delta G(\boldsymbol{R}_A, \boldsymbol{R}_B)$] are referred to as "driving forces," or "thermodynamic driving forces." Indeed, ΔA or ΔG can be used to "predict" in which direction a macroscopic process will occur when a constraint is removed. However, this term applies to macroscopic

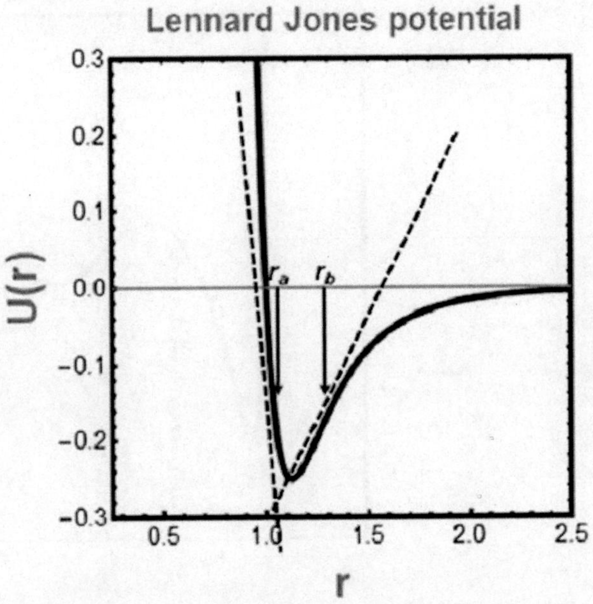

Fig. 3.8 A single potential function *U(r)*. At the two points, r_a and r_b the interaction energies are the same, but the forces are unequal both in magnitude and in sign

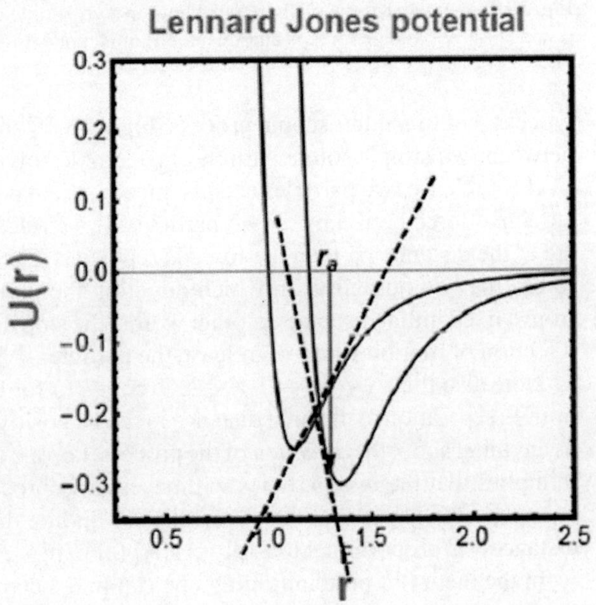

Fig. 3.9 Two potential functions *U(r)*. At a separation r_a, the value of $U(r)$ in one curve is lower than in the other curve, but the forces at the same distance, have different signs; one attractive and one repulsive

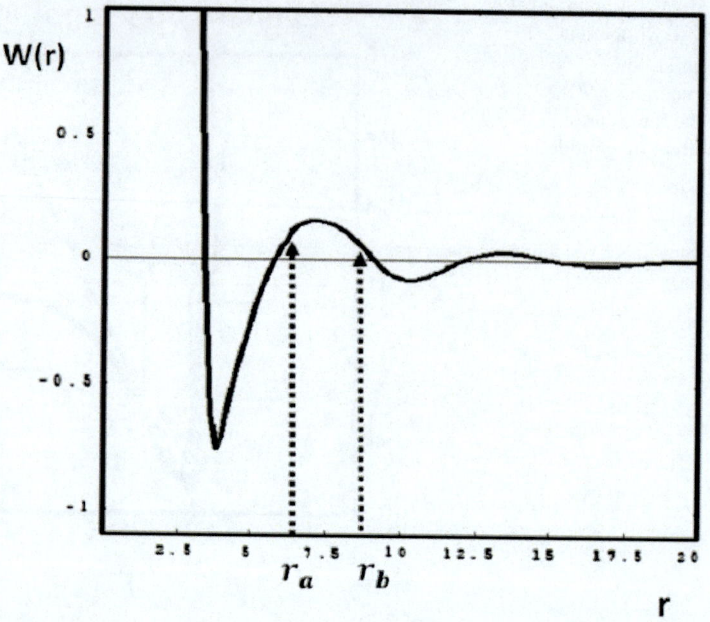

Fig. 3.10 A typical form of the PMF between two simple solute particles in a solvent. At the distance r_a the average force is attractive, at r_b it is repulsive

process, not to a microscopic process. Figure 3.10 shows a typical form of the PMF between two simple solute particles in a simple solvent.

If we release two particles at a distance r_a, there is an over-all (direct and indirect) attractive force between the two particles. If we release the two particles when they are at the distance r_b, then the force is repulsive. These forces are statistical average forces they do not tell us in which direction the particles will move when released from a fixed initial distance. In other words, the slop does not tell us the instantaneous direction of motion when we release the particle.

Note also that $\Delta A(\infty \rightarrow r)$ or $\Delta G(\infty \rightarrow r)$ for bringing the two particles from infinite separation to the final distance r, can be positive or negative. This does not tell us anything about the *direction* of the process, i.e. one cannot say that $\Delta A(\infty \rightarrow r) < 0$ implies that the two particles will necessarily proceed from infinite separation to r. All one can say is that the probability of finding the pair of solute particles at the distance r is proportional to $\exp[-\beta \Delta A]$ (or $\exp[-\beta \Delta G]$ in the T, P, N ensemble).

In the theory of protein folding one can find a common misconception regarding the "energy landscape" of the protein. First, by "energy landscape" one actually means "Gibbs energy landscape." Second, and more importantly people have searched for the "global minimum" in this landscape. This misconception was discussed in detail in Ben-Naim [2], and will not concern us here.

One final comment regarding the meaning of the PMF and the derived mean force. The strength of the interaction as well as the force between the two particles are statistical averages, and these averages are valid only for equilibrium systems.

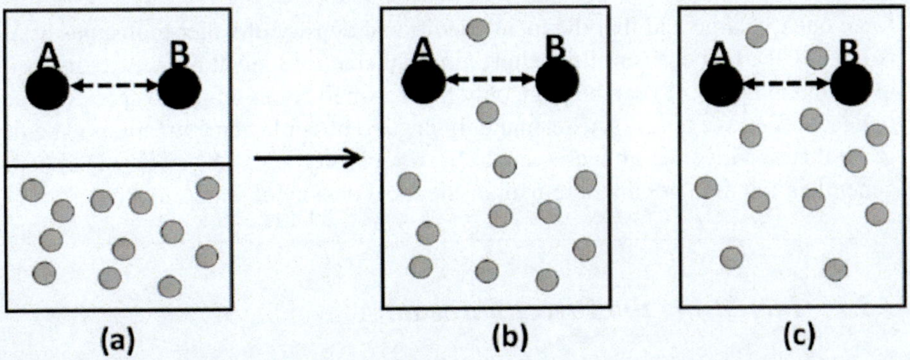

Fig. 3.11 **a** The initial state of a system. The two particles A and B are in Vacuum. The direct force between A and B is indicated by the dashed double arrow. **b** and **c** Two different non-equilibrium micro-states of the system after removing the partition between the two compartments. The direct force between A and B remains the same as in (**a**). The SIF between A and B are different in (**b**) and in (**c**)

Consider again the two particles at $\boldsymbol{R}_A$ and $\boldsymbol{R}_B$ in a "solvent" which is not at equilibrium. Clearly, the *direct* force is independent of the presence of the solvent (this is usually a good approximation). The *indirect* force is an ensemble average, which can be taken either in the T, V, N or T, P, N ensemble. This average is taken over all possible microstates of the system. Therefore, the average must be taken at equilibrium.

Figure 3.11 shows two non-equilibrium states of a system, which results after two short times after removing a partition between two compartments of the system in Fig. 3.11a. Clearly, the direct force between A and B does not change. The indirect force at any particular microstates will be different. In general, we cannot calculate the indirect force at any instant of time. To do that we must know the exact configuration of all the particles. Thus, when we talk about the average indirect force we always mean average over all the possible configurations of the solvent molecules at *equilibrium.*

Similarly, when we refer either to W to ΔA, or to ΔG we mean a difference in a thermodynamic quantity between two *equilibrium* states of the system. The same is true for the *average force*—it is an average over all possible configurations of the solvent molecules at equilibrium.

3.5 Generalization of the Statistical Mechanical Expression for the Indirect Solvent-Induced Force

In this section we generalize Eq. (3.38) in two directions. Before doing so, we recall the meaning of the indirect force as expressed in Eq. (3.38). In order to obtain strong indirect forces we must have a region of integration over $\boldsymbol{X}_W$ in which both the direct force on A is large and that the local conditional density of water molecules at X_W is large. Clearly, because of the short range interactions, each of the factors in the integrand in Eq. (3.38) can be large only from a small configurational space of water molecule X_W. We now discuss separately the two possible generalizations. The first generalization involves an increment of the direct force $-\nabla_A(X_A, X_W)$. The second generalization involves an increment of the local density of water molecules at X_W.

3.5.1 Increasing the Direct Force on A

In deriving Eq. (3.38) we assumed that the two solute particles A and B are $H\phi I$. In Fig. 3.12a we show two segments of either the same protein or of different proteins. The analogue of the case discussed in the previous section is when A is a part of a protein carrying a $H\phi I$ group and B is also a segment of a protein carrying a $H\phi I$ group.

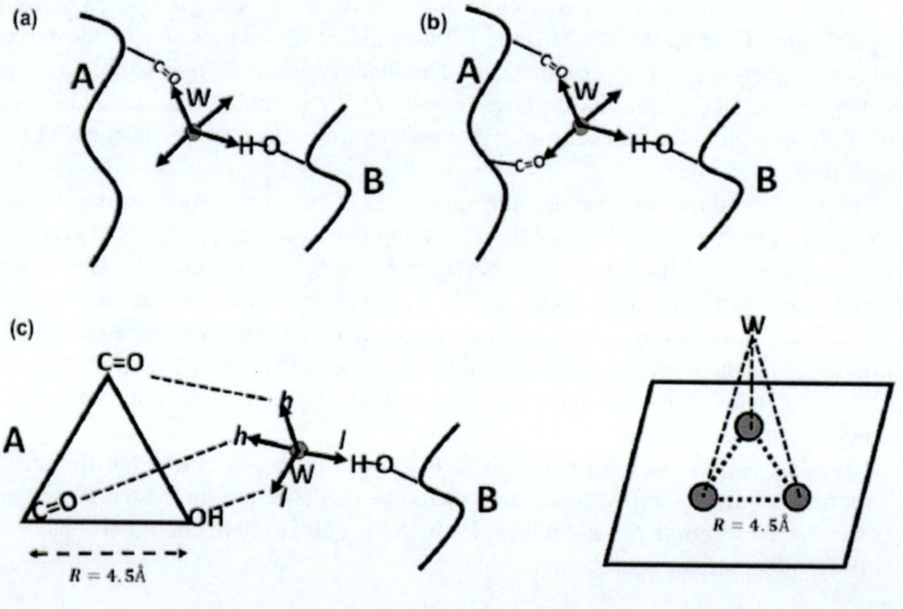

Fig. 3.12 **a** Two segments of a protein, A and B, each carrying a $H\phi I$ group. A water molecule (W), at some favorable configuration, X_W can exert a strong force on A, while B may enhance the local density of water at the configuration X_W. **b** As in (**a**), now A is modified so that carries two $H\phi I$ groups. A water molecule (W), at some favorable configuration, X_W can exert a stronger force on A, while the two $H\phi I$ groups on A may enhance the local density of water at the configuration X_W. **c** As in (**a**), now A is modified so that carries three $H\phi I$ groups. A water molecule (W), at some favorable configuration, X_W can exert a stronger force on A, while the three $H\phi I$ groups on A may enhance the local density of water at the configuration X_W. The letters h and l indicate directions along which a water molecule can form a HB as a donor and acceptor, respectively

To get a strong solvent-induced force on A, we need a relative high local density of water molecules at a specific configuration. This density may be enhanced by a $H\phi I$ group on B, as shown in Fig. 3.12a.

Next, we keep the segment B as in Fig. 3.12a, i.e. it provides an enhanced probability density of water molecule at $\boldsymbol{X}_W$. We now change the segment A so that it can provide two $H\phi I$ groups (say, two hydroxyl groups or two carbonyl groups), in such a configuration that a water molecule at $\boldsymbol{X}_W$ can form two HBs simultaneously with these groups on A, Fig. 3.12b. In such a situation the force $-\nabla_A(\boldsymbol{X}_A, \boldsymbol{X}_W)$ is expected to be stronger, perhaps nearly twice as in the case of Fig. 3.12a. Also, the conditional density at $\boldsymbol{X}_W$ might be larger than that of the previous case, because we have modified the surface of A which was made more $H\phi I$ hence will increase the probability of finding a water molecule at $\boldsymbol{X}_W$, Fig. 3.12b. Clearly, this modification will increase both of the factors in the integrand in Eq. (3.38), i.e. a substantial increase in the SIF on A.

A further increase in the direct force on may be obtained by modifying the surface of A so that it provides *three* $H\phi I$ groups (e.g. two carbonyl and one hydroxyl, or two hydroxyl and one carbonyl) in a triangular configuration such that they can form three HBs with a water molecule at $\boldsymbol{X}_W$, Fig. 3.12c. In this case if the $H\phi I$ group on B is a hydroxyl, then, we choose two carbonyl and one hydroxyl on the segment A. In this case we can expect a stronger direct force exerted by the water molecule on A, as well as a higher density of water molecule at $\boldsymbol{X}_W$ due to the presence of the three groups on the modified A segment, Fig. 3.12c.

In Ben-Naim [3, 4], one can find some estimates of the maximal forces for these cases. Also, an extreme case of a surface that provides four $H\phi I$ groups to form four HBs with a water molecule. We shall not discuss these estimates here because we have no simulated results which confirm the theoretical predictions from 1990. See Chap. 5.

In the three cases shown in Fig. 3.12a–c, we modified the surface of the solute A in such a way that the *direct force* by a water molecule had increased from (a) to (b), to (c). Note that the solvent-induced force was increased because of two factors: First, the modifications in A has increased the *direct force* on A, and second, because of the increase in the local density of water molecules at a favorable configuration to exert such a strong force.

In these three cases we held solute B at a fixed distance and orientation. Therefore, the increment of the force has been achieved for a fixed distance between A and B.

3.5.2 Longer Range Forces Between Two Hydrophilic Surfaces

In the previous subsection we held the distance between A and B, fixed. We increased the force $-\nabla_A(\boldsymbol{R}_A, \boldsymbol{X}_W)$ by modifying the surface of the solute A. We could also modify the surface of the solute B in the same way we did in Sect. 3.4 to enhance

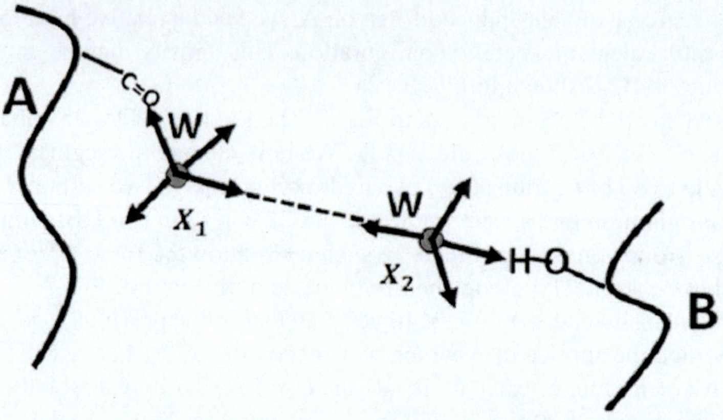

Fig. 3.13 Two segments of a protein each carrying a $H\phi I$ group. The distance between the A and B is larger than in Fig. 3.12. In this case, there is no configuration of a water molecule (W), at which both a strong direct force on A, and an enhance the local density of water. The probability of finding a "two-water-bridge" is very small

the local density $\rho(\boldsymbol{X}_W|\boldsymbol{R}_A, \boldsymbol{R}_B)$. In this case we would further increase the solvent-induced force by increasing both factors of the integrand in Eq. (3.38).

Next, we ask about changing the *range* of the indirect force. Suppose we change the distance between the two solutes A and B, Fig. 3.13. If both A and B provide one $H\phi I$ group then there are no configurations $\boldsymbol{X}_W$ from which both factors in the integrand in Eq. (3.38) will be large.

In this case a water molecule can either be close to the OH group, or it can be close to the carbonyl group. In such case we will have either a strong force $-\nabla_A(\boldsymbol{R}_A, \boldsymbol{X}_W)$ or a high density $\rho(\boldsymbol{X}_W|\boldsymbol{R}_A, \boldsymbol{R}_B)$, respectively, but the product of these two factors cannot be large.

We can however, obtain larger density at $\boldsymbol{X}_W$ by modifying the content of the surface of the solute B. To do this we first modify Eq. (3.38) as follows: Instead of the notation $\boldsymbol{X}_W$, we use the notation $\boldsymbol{X}_1$ for a specific water molecule. We now rewrite the probability density in Eq. (3.38) replacing $\boldsymbol{X}_W$ with $\boldsymbol{X}_1$ as:

$$\begin{aligned}
\rho(\boldsymbol{X}_1|\boldsymbol{R}_A, \boldsymbol{R}_B) &= \frac{\rho(\boldsymbol{X}_1, \boldsymbol{R}_A, \boldsymbol{R}_B)}{\rho(\boldsymbol{R}_A, \boldsymbol{R}_B)} \\
&= \frac{\int \rho(\boldsymbol{X}_1, \boldsymbol{X}_2, \boldsymbol{R}_A, \boldsymbol{R}_B) d\boldsymbol{X}_2}{\rho(\boldsymbol{R}_A, \boldsymbol{R}_B)} \\
&= \frac{\int \rho(\boldsymbol{X}_1, \boldsymbol{X}_2, \boldsymbol{R}_A, \boldsymbol{R}_B) d\boldsymbol{X}_2}{\rho(\boldsymbol{R}_A, \boldsymbol{R}_B)} \times \frac{\rho(\boldsymbol{X}_2, \boldsymbol{R}_A, \boldsymbol{R}_B)}{\rho(\boldsymbol{X}_2, \boldsymbol{R}_A, \boldsymbol{R}_B)} \\
&= \int \rho(\boldsymbol{X}_1|\boldsymbol{X}_2, \boldsymbol{R}_A, \boldsymbol{R}_B) \rho(\boldsymbol{X}_2|\boldsymbol{R}_A, \boldsymbol{R}_B) d\boldsymbol{X}_2 \qquad (3.40)
\end{aligned}$$

In the first step on the right-hand side of (3.40) we used the definition of the conditional probability density as the ratio between the probabilities of the two events

"$(\boldsymbol{X}_1, \boldsymbol{R}_A, \boldsymbol{R}_B)$" and the "$(\boldsymbol{R}_A, \boldsymbol{R}_B)$". In the second step we rewrite the density $\rho(\boldsymbol{X}_1, \boldsymbol{R}_A, \boldsymbol{R}_B)$ as the marginal probability density, derived from the probability density $\rho(\boldsymbol{X}_1, \boldsymbol{X}_2, \boldsymbol{R}_A, \boldsymbol{R}_B)$. The latter is the probability density of finding the event "$(\boldsymbol{X}_1, \boldsymbol{X}_2, \boldsymbol{R}_A, \boldsymbol{R}_B)$." By integrating over all possible values of $\boldsymbol{X}_2$, we obtain the marginal density $\rho(\boldsymbol{X}_1, \boldsymbol{R}_A, \boldsymbol{R}_B)$. Finally, we extend the ratio in the integrand by multiplying and dividing by the same factor $\rho(\boldsymbol{X}_2, \boldsymbol{R}_A, \boldsymbol{R}_B)$ which allows us to use again the definition of the conditional probability to obtain the final result on the right-hand side of (3.40).

The meaning of the integral on the right-hand-side of (3.40) is simple. To obtain the probability of the event "$\boldsymbol{X}_1$" given the condition "$(\boldsymbol{R}_A, \boldsymbol{R}_B)$," we first ask about the probability of the event "$\boldsymbol{X}_1$" given "$(\boldsymbol{X}_2, \boldsymbol{R}_A, \boldsymbol{R}_B)$," *and* the probability of "$\boldsymbol{X}_2$" given "$(\boldsymbol{R}_A, \boldsymbol{R}_B)$," then sum over all possible events "$\boldsymbol{X}_2$." We can now use Eq. (3.40) in Eq. (3.38) to obtain the equation for the indirect force:

$$\boldsymbol{F}_A(\text{indirect}) = \int [-\nabla_A U(\boldsymbol{R}_A, \boldsymbol{X}_1)][\rho(\boldsymbol{X}_1|\boldsymbol{X}_2, \boldsymbol{R}_A, \boldsymbol{R}_B) \times\rho(\boldsymbol{X}_2, \boldsymbol{R}_A, \boldsymbol{R}_B)]d\boldsymbol{X}_1 d\boldsymbol{X}_2 \tag{3.41}$$

As we have pointed out in connection with Fig. 3.13, when we increase the distance between A and B we cannot expect that both of the factors in the integrand in Eq. (3.38) will be large. Equivalently, we cannot expect that the three factors in Eq. (3.41) will be large. However, by modifying the surface of B (in the same way we modified the surface of A in Sect. 3.4 we can get a strong indirect force even at this larger distance between A and B.

First, suppose we have two $H\phi I$ groups in B such that a water molecule can form HB with both groups. In this case the local density at $\rho(\boldsymbol{X}_2|\boldsymbol{R}_A, \boldsymbol{R}_B)$ will be large. As a result the water molecule at $\boldsymbol{X}_2$ can cause an increase in the density at $\boldsymbol{X}_1$ from which the water molecule can exert strong force on A. Thus, by this modification we achieved a strong force by what we referred to originally [3, 4] as a "two-water bridge," Fig. 3.14, mediating between A and B at a distance which is larger than the distance in which we obtain a "one-water bridge."

One can further modify the surface B similar to the manner we modified the surface of A in Fig. 3.12c, to obtain larger local density at $\boldsymbol{X}_2$, hence a larger density at $\boldsymbol{X}_1$ so that the net effect will be a stronger force due to two-water-bridge. Another modification of B is to add three or more hydrophilic groups on B, so as to obtain a "three-water bridge," and obtain strong force at a larger distance between A and B. To do this, we have to repeat the same arguments as we did in Eq. (3.40) but to add another intermediary water molecule. Thus, instead of (3.40) we can write:

$$\rho(\boldsymbol{X}_1|\boldsymbol{R}_A, \boldsymbol{R}_B) = \int \rho(\boldsymbol{X}_1|\boldsymbol{X}_2, \boldsymbol{X}_3, \boldsymbol{R}_A, \boldsymbol{R}_B) \times \rho(\boldsymbol{X}_2|\boldsymbol{X}_3, \boldsymbol{R}_A, \boldsymbol{R}_B)\rho(\boldsymbol{X}_3|\boldsymbol{R}_A, \boldsymbol{R}_B)d\boldsymbol{X}_2 d\boldsymbol{X}_3 \tag{3.42}$$

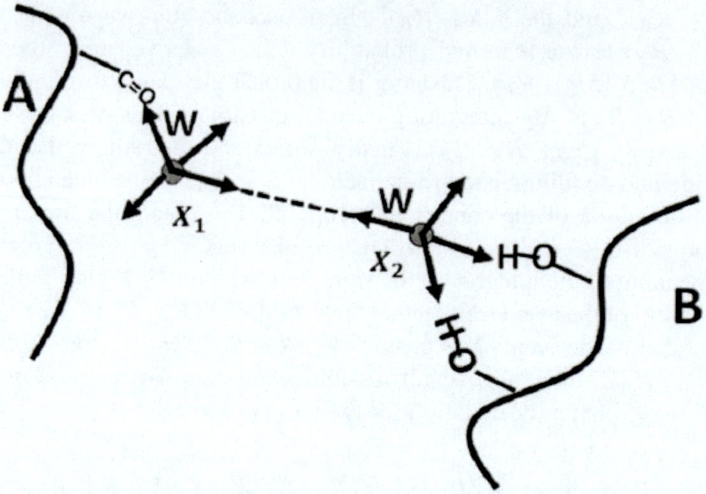

Fig. 3.14 As in Fig. 3.13, now B is modified so that it carries two $H\phi I$ groups. The two $H\phi I$ groups on B may enhance the local density of water at the configuration X_2. This, in turn, may enhance the local density at X_1 from which a strong direct force is exerted on A

Using Eq. (3.42) in Eq. (3.38) we can get an indirect force on A based on "three-water bridge." The rationale for such a strong force, and at a larger distance is similar to the previous case. A specific configuration of three $H\phi I$ groups on B can cause a large density at $\boldsymbol{X}_3$ which can cause a large density at $\boldsymbol{X}_2$, which can cause a large density at $\boldsymbol{X}_1$ from which a strong force may be exerted on A.

The indirect force in this case may be written as:

$$\boldsymbol{F}_A(\text{indirect}) = \int [-\nabla_A U(\boldsymbol{R}_A, \boldsymbol{X}_1)] \times [\rho(\boldsymbol{X}_1|\boldsymbol{X}_2, \boldsymbol{X}_3, \boldsymbol{R}_A, \boldsymbol{R}_B) \times \rho(\boldsymbol{X}_2|\boldsymbol{X}_3, \boldsymbol{R}_A, \boldsymbol{R}_B)\rho(\boldsymbol{X}_3|\boldsymbol{R}_A, \boldsymbol{R}_B)]d\boldsymbol{X}_1 d\boldsymbol{X}_2 d\boldsymbol{X}_3 \quad (3.43)$$

In principle, one could extend this process to four, five, etc. water bridges. However, by doing so we expect that the product of four and more conditional densities will be quite small.

In Ben-Naim [3, 4] and [5], a host of new forces are discussed between plates having $H\phi I$ groups on their surfaces. One can obtain "three-water-bridges", "four-water-bridges," etc. For each kind of bridge, we can have different *strengths* of the direct force, as well as different *range of the interaction between A and B*. For instance, if A contains a pattern of $H\phi I$ groups such that they can induce high local density of water molecules at some distance above the surface of A. This will be equivalent to moving the surface of A towards the surface of B. Examples are discussed in the original article, Ben-Naim [3, 4]. It was also shown that these force can be oscillatory with alternating regions of repulsion and attraction, similar to the PMF between two small solutes, or the observed forces between mica surfaces.

Similar forces are expected to be operating between biological membranes, micelles, or between cells and clay minerals.

3.6 Concluding Remarks

In this chapter we presented the statistical mechanical definitions of the SII and the SIF. It was pointed out that in water, as a solvent one can expect strong SII as well as SIF.

In Chap. 6 we shall discuss a single-water-bridge to explain the self-assembly of proteins. Higher-order bridges might not contribute significantly to the Gibbs energy of association between proteins. However, perhaps such bridges can contribute to the folding and association between segments of DNA. We shall also see in Chap. 6 that the solvent-induced hydrophilic forces are likely the most important factor that determines the speed of the folding process. As such it is safe to say that $H\phi I$ forces "solved" the so-called Levinthal Paradox.

References

1. Ben-Naim A (2006) Molecular theory of solutions. Oxford University Press, Oxford, UK
2. Ben-Naim A (2016) Myths and verities in folding theories. World Scientific, Singapore
3. Ben-Naim A (1990) Solvent-induced forces in protein folding. J Phys Chem 94:6893–6895
4. Ben-Naim A (1990) Strong forces between hydrophilic macromolecules; implications in biological systems. J Chem Phys 93:8196–8210
5. Ben-Naim A (2011) Molecular theory of water and aqueous solutions, part II: the role if water in protein folding, self-assembly and molecular recognition. World Scientific, Singapore

Chapter 4
Hydrophobic Interactions and Forces

As I explained in both the preface and in Chap. 1 of this book, the whole study of the solvent-induced interactions in aqueous solutions was motivated by the protein folding problem and other related processes occurring in aqueous solutions. In this chapter, we shall focus on the SII and forces between $H\phi O$ molecules or groups of atoms attached to a protein. We shall describe some theoretical results, some simulated results, and some experimental results on $H\phi O$ hydrophobic interactions. We also discuss the temperature dependence of the $H\phi O$ interaction in water, which is unique to aqueous solutions.

4.1 Early Evidence for $H\phi O$ Interactions

In this field, when different people say "evidence" of $H\phi O$ interactions, they mean different things. First, there is a common confusion between "interactions" and "forces." Second, in the old literature, there was great confusion between $H\phi O$ interactions and $H\phi O$ solvation. Finally, some people use the fact that most of the $H\phi O$ groups in the globular proteins are in the interior of the protein, as evidence for the importance of the $H\phi O$ interactions in the stability of the 3D structure of the proteins.

What do we mean by "evidence for $H\phi O$ interactions"? Does it mean there is evidence that $H\phi O$ interactions are stronger in water compared to any other liquid, or that there exists evidence that the $H\phi O$ interactions are important (some say "most important" or even dominant) in protein folding?

Prior to the early 1970s, there was no quantitative measure of the strength of $H\phi O$ interactions. In fact, there was not even a proper definition of the $H\phi O$ interactions in terms of well-defined measurable quantities. There were many practitioners in the field of proteins who quoted Kauzmann's idea about the "tendency of non-polar groups to adhere to each other in aqueous solution," as a "definition" of the $H\phi O$ effect. This "tendency to adhere" was implied from the fact that non-polar solutes

A. Ben-Naim, *Solvent-Induced Interactions and Forces in Protein Folding*, Physical Chemistry in Action, https://doi.org/10.1007/978-3-031-31871-9_4

Table 4.1 Values of the standard Gibbs energy of dimerization for three carboxylic acids in water at 25 °C

Compound	$\Delta G°$/(kcal/mol)
CH_3-COOH	– 0.8
CH_3CH_2-COOH	– 1.1
$CH_3CH_2CH_2-COOH$	– 1.4

are sparingly soluble in water, hence, these molecules tend to avoid contact with water—hence, the term "hydrophobic". When two non-polar groups or molecules are brought into contact-distance between them, they reduce their exposure to water. In fact, many authors viewed the "tendency to adhere to each other" as a partial reversal of solubility. All these hand-waving arguments were qualitative and could not serve as a satisfactory definition of $H\phi O$ interactions.

Until the 1970s there were only two pieces of "evidence" for the "tendency to adhere to each other." One is the Gibbs energy of dimerization, as suggested by Schrier et al. [1] see Table 4.1.

The second experimental quantity relevant to the problem of $H\varphi O$ is the second virial coefficient B_2^* in the density expansion of the osmotic pressure, suggested by Kozak et al. [2]. These two quantities show that when you increase the $H\phi O$ chain in a series of alcohols or carboxylic acids, the overall interaction between the two molecule increases with the chain length. Unfortunately, these measures were very qualitative measures of the "tendency to adhere to each other."

In fact, both measures involve molecules containing both polar and non-polar groups. Therefore, it is difficult to separate between the $H\phi O$ and the $H\phi I$ contributions. There existed no measure of the $H\phi O$ interaction involving only two non-polar groups or molecules. The explanation of the increasing (in absolute value) of the Gibbs energy of dimerization of carboxylic acids, was based on the assumption that there are two possible "structures" of a dimer, as shown in Fig. 4.1. Only in one of these structures, the two $H\phi O$ groups are close enough to contribute to the total interaction energy.

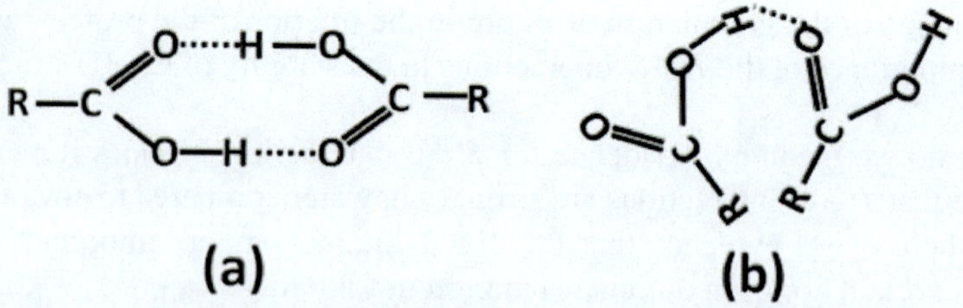

Fig. 4.1 Two possible structures of the dicarboxylic acid dimer. In one **a** there are two HBs between the monomers, but the distance between the $H\phi O$ is large. In the second **b** there is one HB, but the distance between the $H\phi O$ is relatively small

In the early 1970s, a new definition of $H\phi O$ was proposed in terms of the *indirect work* involved in the process of bringing two non-polar solutes from infinite separation to contact distance. This definition was discussed in Chap. 3, and in more detail in my book Ben-Naim [3].

Interestingly, the very definition of the $H\phi O$ interaction led to a discovery of a useful measure of a "pure" $H\phi O$ interaction, which I will describe later in this chapter, and in more detail in Ben-Naim [3].

The definition that I suggested in the early 1970s, was based originally on the process of bringing two $H\phi O$ simple particles, such as argon atoms or even hard-spheres, from infinite separation to a distance *R* within the liquid at constant P, T, N_w. The Gibbs energy change for this process is:

$$\begin{aligned}\Delta G(\infty \to R) &= G(\infty) - G(R)\\ &= U(R) + \delta G(R)\end{aligned} \tag{4.1}$$

Since the direct interaction $U(R)$ is supposed to be independent of the solvent and is the same as the interaction between the same two particles in a vacuum. I suggested referring to $\delta G(R)$ as the $H\phi O$ interactions. Originally, when the only SII discussed in connection with proteins was the $H\phi O$, I used the short-term notation HI for "hydrophobic interaction." Later, I had to distinguish between hydrophobic ($H\phi O$) and hydrophilic ($H\phi I$) interactions.

Today, when one writes the process ($\infty \to R$) in a vacuum and in a liquid phase, as in Fig. 1.7, one can immediately see that $\delta G(R)$ is the difference between the solvation Gibbs energies of the two separate solutes at infinite separation, and the pair of solute at distance *R*, i.e.

$$\delta G = \Delta G_D^*(R) - 2\Delta G_M^* \tag{4.2}$$

This relationship could not be obtained from the old definition of the solvation Gibbs energies [for details, see reference [4], Ben-Naim (1987)].

Note that neither the dimerization constant, nor the second virial coefficient B_2^* could be measured for the "dimer" denoted D in Eq. (4.2). Additionally, there was no way to measure the solvation Gibbs energy of this hypothetical dimer D.

In the early 1970s. I had to work out the statistical mechanical expression for δG using the so-called "*charging process*" of a solute in a solvent. See Ben-Naim [5]. The result of this long derivation was already suggestive for an approximate measure of the $H\phi O$ interaction.

The statistical mechanical expression for δG showed clearly that it does not depend on the *direct* interaction potential function $U(R)$. Therefore, it was argued that Eq. (4.2) could be applied to any distance *R*, not necessarily distances which are larger than the diameter of the solute particles σ. In fact, Eq. (4.2) could be applied even for $R = 0$, i.e. for the process of bringing two solute particles from infinite separation to zero equation, Fig. 4.2. Hence, Eq. (4.2) could be written for this process as:

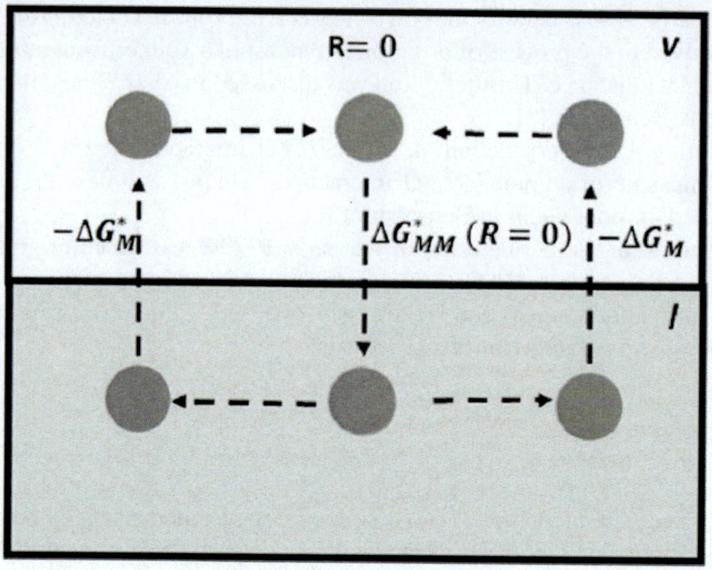

Fig. 4.2 A cyclic process, similar to the one in Fig. 1.7. Instead of a contact distance in Fig. 1.7, we bring here two simple monomers (M) to the final distance of R = 0

$$\delta G(\infty \to 0) = \Delta G^*_D(R = 0) - 2\Delta G^*_M \tag{4.3}$$

where $\Delta G^*_D(R = 0)$ is the solvation Gibbs energy of a "dimer" of solute particles at zero distance.

For some time, I entertained the idea of replacing this hypothetical dimer with a *real* solute in water that interacts with water as two-solute particles at zero separation.

The simplest solute particle that one can think of is a hard sphere particle of diameter σ. When brought to a distance of $R < \sigma$, two such particles will feel infinite repulsive forces, i.e. $U(R)$ tends to infinity when R approaches σ.

However, in the definition of $\delta G(R)$, we have already excluded the direct pair potential. This means that δG is the Gibbs energy change for bringing two *cavities* from infinite separation to zero separation. Of course, two cavities do not repel each other as two hard particles. Furthermore, the "dimer" D in this case (i.e. two cavities at zero separation) interacts with water molecules exactly like one cavity at some fixed position.

Thus, for two hard-sphere particles at zero separation, one can rewrite Eq. (4.3) as:

$$\delta G_{HS}(\infty \to 0) = \Delta G^*_{HS} - \Delta G^*_{HS} - \Delta G^*_{HS} = -\Delta G^*_{HS} \tag{4.4}$$

This means that the Gibbs energy change for bringing two cavities from infinite separation to zero separation is simply the Gibbs energy of *desolvation* of one cavity, or equivalently, the desolvation of a hard sphere solute in a solvent. This could serve as a measure of "pure" $H\phi O$ interaction. This result led me to apply the scaled particle theory (SPT) to the problem of $H\phi O$ interaction, a topic that is fully discussed in Ben-Naim [3].

A more practical measure of $H\phi O$ interaction based on solvation Gibbs energies of real solute particles was developed which turned out to be quite useful in comparing the SII between two real non-polar solutes in different solvents.

In the following we discuss very briefly the derivation of the new measure of $H\phi O$, and present a sample of results. More details can be found in Ben-Naim [3].

In Chap. 3, we showed that the Gibbs energy change for the process of bringing two solute particles from infinite separation to any final distance R may be written as:

$$\begin{aligned}\delta G(\infty \to R) &= \Delta G_D^*(R) - 2\Delta G_M^* \\ &= -kT \ln \frac{\langle \exp[-\beta B_D(R)] \rangle_0}{\langle \exp[-\beta B_M] \rangle_0^2}\end{aligned} \tag{4.5}$$

In Eq. (4.5), $B_D(R)$ is the total binding energy of the pair of solute molecules (referred to as the "dimer") at a distance R, and B_M is the binding energy of the monomer (M). [When $r = 0$, the numerator will be equal to one factor in the denominator). The average denoted $\langle \rangle_0$ is over all the configurations of the solvent molecules (excluding the solute molecules] in the T, P, N ensemble.

Now suppose that the solute molecules are methane. This is the simplest $H\phi O$ molecule that interacts with water similar to a methyl group in proteins. In reality, we can bring two methane molecules to a distance of "contact" which is about $R \approx 3.82$ Å.

Unfortunately, we cannot measure the solvation Gibbs energy of such a "dimer." We note, however, that in Eq. (4.5) the methane-methane pair potential does not feature. This means that we do not need to apply the equation to *real* methane molecules, but only two "fields of force" which are equivalent to two methane molecules. All we need is that the binding energy, $B_D(R)$ in Eq. (4.5) be exactly the same as for two methane molecules. The "fields of force" are like two virtual molecules that have the same *binding energy* as two real methane molecules, but which do not interact with each other. Once we realize that the direct solute–solute interaction does not feature in Eq. (4.5), we can choose in Eq. (4.5) any distance R we wish (including $R = 0$).

A particular choice of distance, which led me to create a useful measure of the strength of the $H\phi O$ interaction, is $R = \sigma_1 = 1.53$ Å. This is the carbon–carbon equilibrium distance in an *ethane* molecule. When we bring two *methane* molecules from infinite separation to a distance $\sigma_1 = 1.53$ Å, the SII (in terms of Gibbs energy change) is $\delta G(\infty \to R)$ in Eq. (4.5). From the point of view of the solvent, the binding energy of this "dimer" at $R = \sigma_1$ is almost the same as the binding energy of a single ethane molecule. Once we realize that, we can "fool the solvent" by replacing the two *methane* molecules at σ_1 by one *ethane* molecule we can write Eq. (4.5) as:

$$\delta G(\infty \to R) = \Delta G^*_{Et} - 2\Delta G^*_{Me} \quad (4.6)$$

At this stage, we have two experimentally measurable quantities on the right-hand-side of Eq. (4.6); the solvation Gibbs energies of methane and ethane molecules.

The quantity $\delta G(\sigma_1)$ is thus the indirect work (or the solvent-induced Gibbs energy change) associated with the process of bringing two methane molecules from an infinite separation to the final distance (σ_1), see Fig. 4.3. This is therefore an approximate measure of "pure" $H\phi O$interaction. Admittedly, this is not a measure of the $H\phi O$ interaction at a realistic distance. However, if we are interested in comparing the strength of the SII between two $H\phi O$ molecules in different solvents this measure turned out to be a very useful one.

Figure 4.4 shows the values of $\delta G(\sigma_1)$ as defined in Eq. (4.6) for water and in a few alcohols. There are two prominent differences between the values in water and in these alcohols. First, $\delta G(\sigma_1)$ in water is much larger (in absolute magnitude) in comparison with the values in alcohols. Second, the temperature dependence of $\delta G(\sigma_1)$ in water has a clear-cut negative slope in water, whereas in other liquids the temperature dependence is very small, sometimes within the experimental error.

Figure 4.4 established the uniqueness of the SII behavior between two $H\phi O$ solute molecules in water compared to all other solvents for which the relevant data were available. In Ben-Naim one can find a great deal of information on $\delta G(\sigma_1)$ in different solvents, and also similar data on the SII among three or more solute molecules. We note here that the temperature dependence of the SII in water is

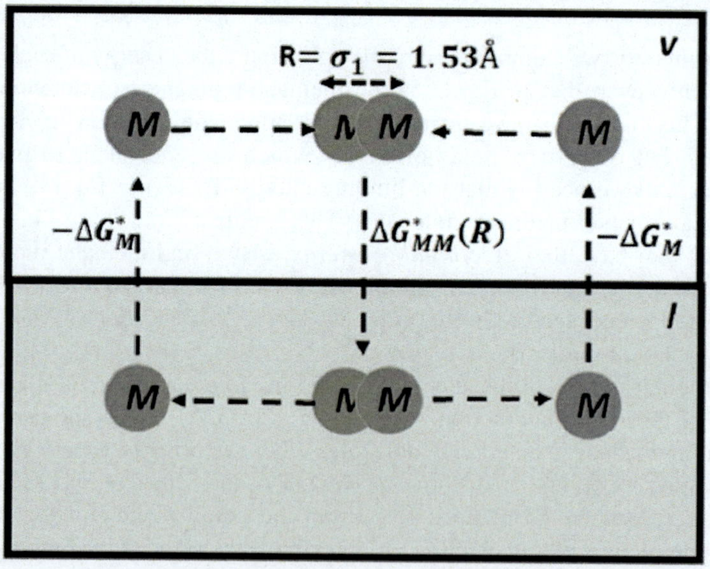

Fig. 4.3 A cyclic process, similar to the one in Fig. 1.7. Instead of a contact distance in Fig. 1.7, we bring here two methane molecules (M) to the final distance of R = σ_1 = 1.53 Å

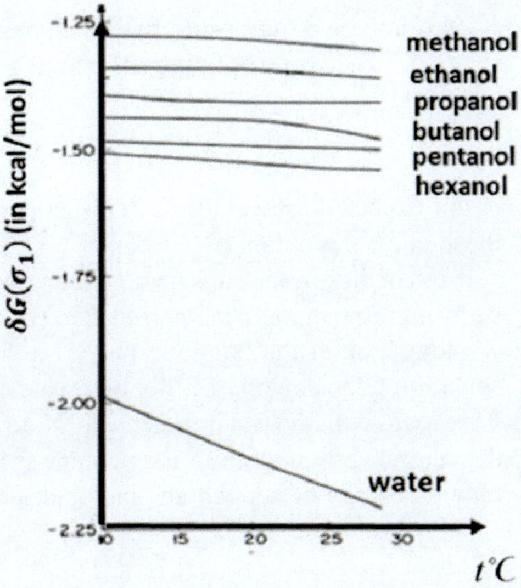

Fig. 4.4 Some values of $\delta G(\sigma_1)$ in water and in some linear alcohols as a function of temperature

related to the change in the "structure" of water. This topic is briefly discussed in Sect. 4.3.

In the rest of this section we discuss a few results on the SII between more than two non-polar particles. In general, using the same type of approximation made in Eq. (4.6) we could define the SII associated with the process of bringing three or more methane molecules from infinite separation to the final configuration of one propane, butane, etc. molecule.

As expected the SII (δG) between three methane groups at the final configuration of propane is stronger (more negative) than for two methane molecules. Likewise, the SII behavior between four methane molecules at the final configuration of the linear butane is even stronger than the SII for the three methane molecules at the configuration of propane. Thus, it was found that:

$$|\delta G_2(\text{ethane})| < |\delta G_3(\text{propane})| < |\delta G(n\text{-butane})| \tag{4.7}$$

Also, it was found that the temperature dependence of all the SII of $H\phi O$ molecules in water is sharply negative (i.e. the strength of the SII increases with temperature). The SII for the same processes in non-aqueous solvents, is almost temperature independent. For more details, see Ben-Naim [3], this chapter.

We mention here one important generalization, as well as an improvement on the approximation involves in Eq. (4.6), for the process of bringing four methane molecules to two different configurations.

Suppose that we can bring four methane molecules to form n-butane, the corresponding SII is:

$$\delta G_4(n\text{-butane}) = \Delta G^*_{n\text{-butane}} - 4\Delta G^*_{n\text{-methane}} \tag{4.8}$$

Similarly, we may bring five methane molecules to form neopentane molecule, Fig. 4.6a. The corresponding SII is:

$$\delta G_5(\text{neopentane}) = \Delta G^*_{\text{neopentane}} - 5\Delta G^*_{\text{methane}} \tag{4.9}$$

As expected the values of δG_5(neopentane) are lower (i.e. more negative) than those of δG_4(n-butane).

Both of these quantities involve the same approximation we have done in Eq. (4.6), i.e. bringing two methane particles to for an ethane molecule.

Now, look at the configuration of a single neopentane molecule, Fig. 4.5. The solvation Gibbs energy of this molecule depends on the total binding energy of all *five* groups with the water molecules. Note that from the "point of view of the solvent" the central carbon atom of neopentane is almost "unseen." In other words, the total binding energy of neopentane molecules can be approximated by:

$$B(\text{neopentane}) = \sum_{i=1}^{5}\sum_{j=1}^{N} U(\mathbf{R}_i, \mathbf{X}_j) \approx \sum_{i=2}^{5}\sum_{j=1}^{N} U(\mathbf{R}_i, \mathbf{X}_j) \tag{4.10}$$

In (4.10) we eliminated the binding energy of the central carbon atom (numbered 1), i.e. we assumed that the binding energy of the central carbon atom is negligible:

$$\sum_{j=1}^{N} U(\mathbf{R}_1, \mathbf{X}_j) \approx 0 \tag{4.11}$$

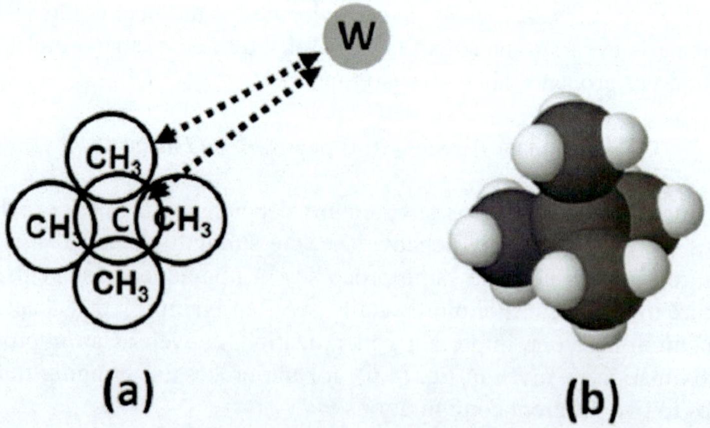

Fig. 4.5 Schematic **a** and a space-filling model **b** of neopentane. The central carbon atom is shielded by the four peripheral methyl groups. Therefore, the interaction between the central atom and water molecules may be neglected in calculating the binding energy of the neopentane molecule

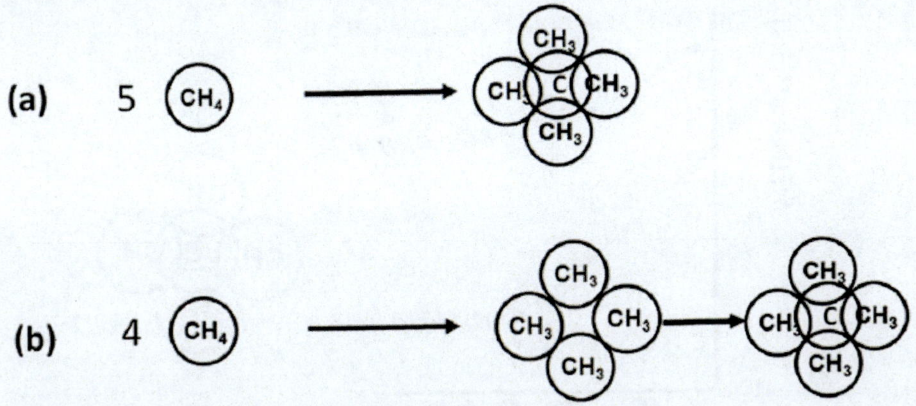

Fig. 4.6 Two processes for which δG may be estimated. **a** Bringing 5 methane (m) molecules to the configuration of neopentane. **b** Bringing 4 methane (m) molecules to the configuration of neopentane

This approximation follows from the fact that the distance between the central carbon and any water molecule is very large, therefore it does not contribute to the total binding energy of the neopentane molecule.

Using this approximation, we can now define the SII interaction among *four* methane molecules at the final configuration as the four peripheral methyl groups shown in Fig. 4.6b. We call this configuration of the four methane molecules the "aggregate" and write the SII as:

$$\delta G_4(agg.) = \Delta G^*(\text{neopentane}) - 4\Delta G^*(\text{methane}) \tag{4.12}$$

Compare Eq. (4.12) with Eq. (4.9). We realize that in Eq. (4.12) we bring the four methane molecules to the more *realistic* configuration than in the previous approximations. The values of $\delta G_4(agg.)$ in (4.12) turn out to be smaller (in absolute values) than the values of $\delta G_4(\text{butane})$ and the latter smaller than the values of $\delta G_5(\text{neopentane})$, Fig. 4.7. The theoretical reason for this is discussed in the next section. More details on the data are in Ben-Naim [3, 6].

4.2 Some Theoretical Results About the $H\phi O$ Interactions

We start by saying that there is no general *theory* of $H\phi O$ interactions. Some people refer to the article of Pratt and Chandler entitled "Theory of hydrophobic effect" as a paper on the "*Theory of Hydrophobic Interaction.*" Unfortunately, the work of Pratt and Chandler [7], did not offer any "theory" of $H\phi O$ interaction. In their article, Pratt and Chandler tried to solve the Ornstein–Zernike (OZ) equations for aqueous solutions of non-polar solutes. In fact, they did not solve the OZ equations

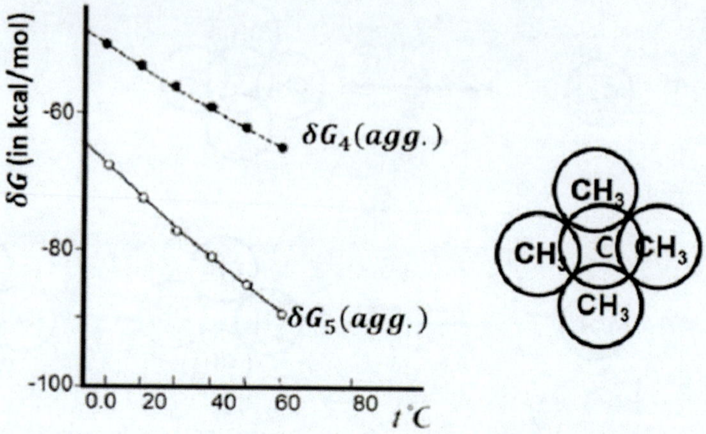

Fig. 4.7 Some values of δG for bringing 4 and 5 methane molecules to the configuration of neopentane, as a function of temperature

for water and a non-polar solute. Instead of using the full (angular dependent) pair correlation function for the water, they use the experimental *spatial pair correlation* function, which depends only on the water-water distance and not on the orientations of the water molecules. By doing so, they effectively treated water molecules as being spherical particles, interacting through a pair potential depending only on the distance, hence, missing the peculiarities of the water-water interactions. See also Sect. 4.3. Therefore, although the article by Pratt and Chandler [7] is titled "*Theory of Hydrophobic Interaction*," it contains no theory of hydrophobic interaction.

In this section, we discuss a few theoretical arguments which are relevant to the SII between two simple non-polar solute molecules in water.

In the previous section, we noted that the measure of the strength of the SII was designed based on the fact that $\delta G(R)$ is independent of the solute–solute pair potential. Therefore, we could replace the two solute particles by two "fields-of-forces," which in the case of the two hard-sphere particles, these "fields of forces" are equivalent to two cavities brought to any distance r including $R = 0$.

Recall the relationship between the PMF and the pair correlation function:

$$g(R) = -kT \ln W(R) = -kT \ln U(R) - kT \ln \delta G(R) \tag{4.13}$$

This equation suggests the writing of the pair correlation function as a product of two functions:

$$g(R) = y(R) \exp[-\beta U(R)] \tag{4.14}$$

where we defined:

$$y(R) = \exp[-\delta G(R)] \tag{4.15}$$

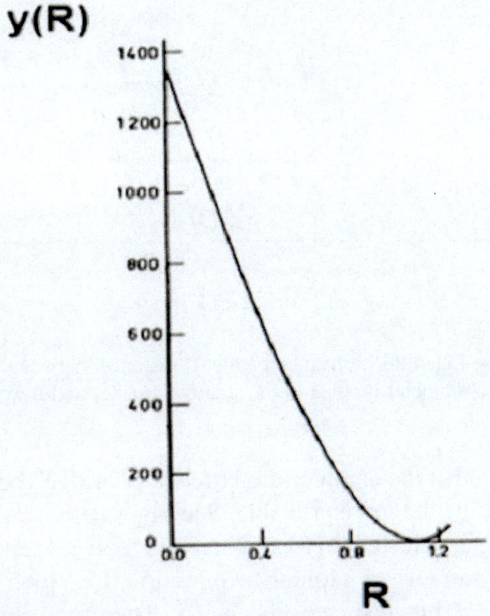

Fig. 4.8 The general form of the indirect part of the pair correlation function

The factor $y(R)$ is the indirect part of the pair correlation function.

Figure 4.8 shows the general form of the function $y(R)$ for two hard-sphere particles of a diameter σ in a hard sphere fluid. There is also a theoretical argument showing that $y(R)$ is a monotonically decreasing function of R in the region $0 \leq R \leq \sigma$ for two cavities in any solvent. This behavior is related to the monotonically decreasing excluded solute of the two hard particles when they are brought to distances smaller than σ. See Ben-Naim [5].

Thus, we can estimate that for $R = 0$, for two methane molecules the value of $y(R = 0)$ is always larger than the value at $y(R = \sigma_1 \approx 1.53\ \text{Å})$, and that the value of $y(R = \sigma_1)$ is smaller than that of $y(R = \sigma)$, i.e. at contact distance.

The second theoretical argument is about the value of $y(R)$, or $\delta G(R)$ at distances larger than σ. For simple spherical particles it is known that $g(R)$ has a maximum at about $R = 2\sigma$. This is sometimes referred to as the solvent-separated distance.

When the diameter of the solute particles is the same as that of the solvent particles, the maximum in $g(R)$ (as well as in $y(R)$ is at 2σ). When the diameter of the solvent particle is different, say σ_{sol}, then the maximum will occur at the distance of $(\sigma + \sigma_{sol})$, Fig. 4.9.

The "solvent-separated" argument is valid for simple spherical particles, see Fig. 4.9. It is in general invalid for aqueous solutions, or for the $H\phi O$ interaction. In fact, some early theoretical calculations by Pratt and Chandler [7], show a maximum in $y(R)$ (or a minimum in $\delta G(R)$) at about 7 Å which is roughly the configuration of "solvent separated" distance between two methane molecules in water. See Fig. 4.10a. However, a simple theoretical argument leads to the prediction

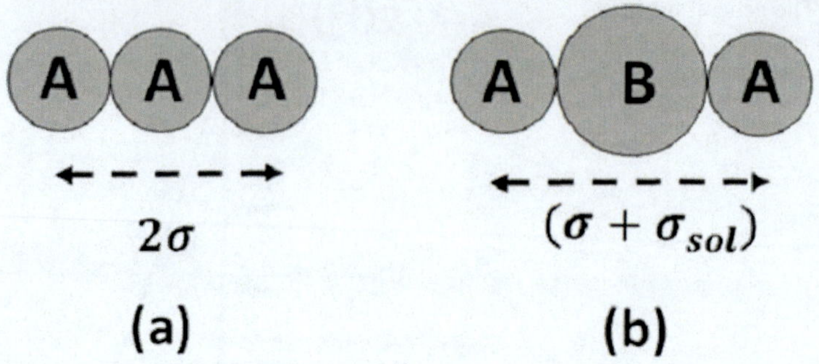

Fig. 4.9 "Solvent-separated" distance between two A solute molecules, when **a** the solvent diameter is equal to the solute diameter, and **b** when the diameters are different

that the configuration of "water bridge" between the two methane molecules should be the more probable. See Fig. 4.10b.

The reason is simple in the "solvent separated" configuration one direction (either an OH or a lone-pair direction, L), along which a water molecule can form a HB, is blocked, see Fig. 4.10a. Therefore, this configuration is less probable. On the other hand, the configuration for which there is no blocking of the hydrogen bonding capability will be more favorable, as in Fig. 4.10b. The distance between the two methane molecules at this configuration is about 6 Å Fig. 4.10b.

In an article on the "Theory of hydrophobic effect," Pratt and Chandler [7] attempted to calculate the function $\delta A(R)$. They used what they called the "Ornstein–Zernike-like" equations in which the spatial water-water pair correlation was used instead of the angle-dependent pair correlation function. Although they could obtain agreement with the experimental value of $\delta G(\sigma_1)$ as defined in Eq. (4.6) they failed

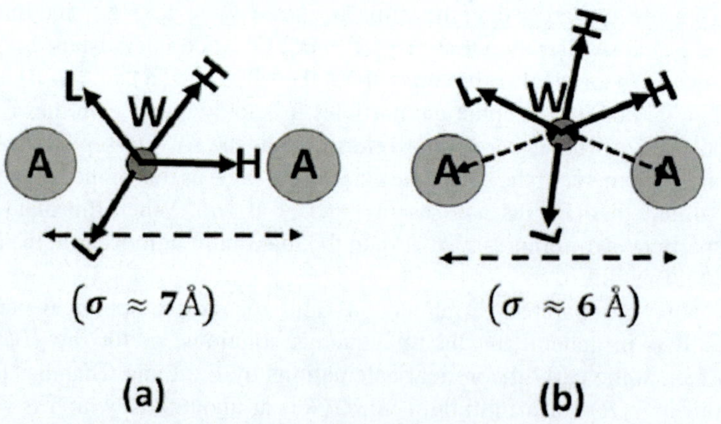

Fig. 4.10 Two methane molecules at **a** water-separated distance, and **b** at water-bridge distance. The bridge is shown by two dashed lines connecting A to W and W to A

to obtain the correct minimum of $\delta A(R)$ or $\delta G(R)$ at the "solvent bridged" configuration. More importantly, they failed to obtain the correct temperature dependence of the $H\phi O$ interaction. This topic will be discussed in Sect. 4.3.

Another theoretical approach to the $H\phi O$ interactions is based on the scaled-particle theory (SPT).

The SPT was originally developed for the study of hard-sphere liquids (see Reiss [8], and some comments about the SPT in Ben-Naim [9]). As such it was a very successful theory. It was also applied in the study of aqueous solutions. The first successful application of SPT to calculate the solvation thermodynamic quantities of argon in various liquids was published by Pierotti [10, 11].

It should be noted however that the SPT is not a "pure" statistical mechanical theory of liquids in the following sense. A statistical mechanical theory uses the molecular properties of molecules as "input" to predict thermodynamic quantities such as the Gibbs or Helmholtz energy, as "output". If we were to denote by *mp* all the molecular properties of say, water molecules, we should be able to calculate the function $A(T, V, N; mp)$, or the Gibbs energy function $G(T, P, N; mp)$. Thus, the "input" in the theory is the molecular properties *mp*, and the "output" of the theory is the thermodynamic quantities (attainable from either from the function $A(T, V, N)$ or $G(T, P, N)$.

In the SPT, the only molecular property (*mp*) used is the *diameter* of the water molecule σ, which also determines the effective hard-sphere interaction potential function. In addition, one uses the *density* of the solvent as well as its temperature dependence, as *input* in the theory. For instance, in liquid water, the temperature dependence of the density is known to be anomalous. By using the density as "input" into the theory one actually already uses macroscopic information which should be obtained as "output," not "input" of the theory. This is a reason why the SPT was quite successful even though it uses only the diameter of the molecules as molecular parameters.

In Sect. 4.6 we shall report some results for the $H\phi O$ interaction among *n* hard-spheres in water to form a spherical aggregate in water.

4.3 Temperature Dependence of the $H\phi O$ Interactions

Formally, the temperature dependence of the SII may be obtained from Eq. (4.5). Taking the derivative of $\delta G(R)$ with respect to T, and constant pressure P, we get:

$$\begin{aligned}\frac{\partial \delta G(R)}{\partial T} &= \frac{\partial \Delta G_P^*(R)}{\partial T} - 2\frac{\partial \Delta G_M^*(R)}{\partial T} \\ &= -\Delta S_D^*(R) + 2\Delta S_M^* \end{aligned} \quad (4.16)$$

In this equation, M denotes the "Monomer," and D denotes the "dimer", i.e. two monomers at a distance *R*.

As we have seen in Sect. 4.1, the temperature dependence of $\delta G(R)$ is negative, i.e. $\delta G(R)$ becomes larger (in absolute magnitude) as we increase the temperature. This phenomenon is quite unique to aqueous solutions.

There are several ways of writing the statistical mechanical expression for the solvation entropy. The simplest one is to take the temperature derivative of either ΔG_M^* (at constant pressure) or ΔA_M^* (at constant volume) to get:

$$\Delta S_M^* = -\left(\frac{\partial \Delta A_M^*}{\partial T}\right)_{T,V} = k \ln\langle \exp[-\beta B_M]\rangle_0 + \frac{1}{T}[\langle B_M\rangle_M + \langle U_N\rangle_M - \langle U_N\rangle_0] \tag{4.17}$$

A similar expression may be obtained for the derivative of ΔG_D^* (see Ben-Naim [6], Chap. 3). In Eq. (4.17) $\langle B_M\rangle_M$ is the average binding energy of M to the solvent molecules, averaged over all configurations of the solvent molecules with the conditional distribution:

$$P(\boldsymbol{X}^N|\boldsymbol{R}_M) = \frac{\exp[-\beta U_N(\boldsymbol{X}^N)]\exp[-\beta B_M(\boldsymbol{R}_M, \boldsymbol{X}^N)]}{\int \exp[-\beta U_N(\boldsymbol{X}^N) - \beta B_M(\boldsymbol{R}_M, \boldsymbol{X}^N)]d\boldsymbol{X}^N} \tag{4.18}$$

$\langle U_N\rangle_M$ is the conditional average of the total interaction energy among the water molecules, and $\langle U_N\rangle_0$ is the (unconditional) average over all configurations of the pure solvent molecules, i.e. in the absence of the solute.

The difference $\langle U_N\rangle_M - \langle U_N\rangle_0$ may be interpreted as the "structural changes" induced in the solvent, followed by the insertion of a solute at a fixed position. For details see Ben-Naim [6].

It is well known that the solvation entropy of nonpolar solutes in water is large and negative. This fact is interpreted in terms of "structural changes" induced in the solvent. See Ben-Naim [6, 12]. Thus, the temperature dependence of the $H\phi O$ interaction is also interpreted as due to a large contribution due to structural changes in the solvent (for the process of bringing the two solute particles from infinite separation to the final configuration.

Recall that $\delta G(R)$ may be written as:

$$\delta G(R) = -kT \ln y(R) \tag{4.19}$$

where $y(R)$ is the indirect part of the pair correlation function. In general, $\delta G(R)$ would depend on temperature, explicitly through T in Eq. 4.19 and implicitly through $y(R)$. Curiously, Pratt and Chandler [7] claimed that the temperature dependence of $\delta G(R)$ is "trivial" and is due to the factor kT in Eq. (4.19). This is not true. The temperature dependence of $\delta G(R)$ has two terms, which we can write as:

$$\frac{\partial G(R)}{\partial T} = -k \ln y(R) - kT\frac{\partial \ln y(R)}{\partial T} \tag{4.20}$$

The second term on the right-hand side of (4.20) includes all contributions due to structural changes in the solvent induced by the process we denoted as ($\infty \rightarrow R$) in Eq. (4.5). Regarding the overall large and negative temperature dependence of $\delta G(R)$, Pratt and Chandler [7] wrote:

> Viewed on a microscopic level this conclusion appears paradoxical. If the hydrophobic interaction is a result of the peculiar nature of liquid water, then one would expect that the hydrophobic interaction strength should decrease as the temperature is raised and liquid water becomes less peculiar.

This is a faulty argument appears frequently in the literature in connection with the temperature dependence of the solvation Gibbs energy of non-polar solutes in water, as well as the Gibbs energy of transfer from water to heavy water. While it is true that both the entropy of solvation of simple non-polar solutes and the difference in the entropies of solvation in Eq. (4.16) are due to the "peculiar nature of liquid water," it is not true that these quantities should change monotonically with the "extent of peculiar properties of water (i.e. when the temperature decreases, or when we move from H_2O to D_2O, the peculiar properties do not necessarily change in the same direction).

In Ben-Naim [6], a detailed argument on the molecular reasons for the structural changes induced in water, is provided. Here, we provide only a qualitative argument as to why we expect that the right-hand side of Eq. (4.16) should be negative. ΔS_M^* is known to be large and negative for non-polar solutes in water (due to structural changes in the water). We expect that both ΔS_M^* and ΔS_D^* to be of the same order of magnitude. Hence, the difference on the right-hand side of Eq. (4.16) is expected to be of the order of ΔS_M^*, i.e. large and negative. In other words, from the fact that the solvent is more structured, it does not follow the $H\phi O$ interaction should be larger. The fact that Pratt and Chandler [7] got the wrong temperature dependence of the $H\phi O$ does not render the experimental facts "paradoxical"; it only proves that the approximations used in their calculations are inadequate.

In terms of structural changes induced in the solvent we can say that the surroundings of the solute M are more structured than that of a pure water. When we bring the two solute molecules from infinite separation to a distance σ, the overall structure around the "dimer" will be less than the sum of the structure around the two separate solute molecules. Of course, this is a very qualitative argument. A more quantitative argument would require a definition of the concept of the "structure of water." See Ben-Naim [6, 12].

Another interesting interpretation of the temperature derivative of the SII is in terms of the Shannon measure of information (SMI). In Eq. (4.17) we wrote the solvation entropy in a way that suggests an interpretation in terms of structural changes in the solvent. We now derive a new expression for the solvation entropy. We start with the configuration PF of a system of one solute M and N solvent molecules.

$$Z_{N+1} = \int \exp[-\beta B_M - \beta U_N] d\mathbf{R}_M d\mathbf{X}^N$$

$$= V \int \exp[-\beta B_M - \beta U_N] d\boldsymbol{X}^N \tag{4.21}$$

Note that the integration is over all the locations of the solute M produce the volume V.

We now rewrite Eq. (4.17) as:

$$\Delta S_M^* = -k \int Pr(\boldsymbol{X}^N|\boldsymbol{R}_M) \ln Pr(\boldsymbol{X}^N|\boldsymbol{R}_M) d\boldsymbol{X}^N$$
$$+ k \int Pr(\boldsymbol{X}^N) \ln Pr(\boldsymbol{X}^N) d\boldsymbol{X}^N \tag{4.22}$$

Here, the conditional probability density $Pr(\boldsymbol{X}^N|\boldsymbol{R}_M)$ is the same as defined in Eq. (4.18).

In Eq. (4.22) the two integrals have the structure of the SMI associated with the probability distributions $Pr(\boldsymbol{X}^N|\boldsymbol{R}_M)$ and $Pr(\boldsymbol{X}^N)$. The SMI is defined for any distribution $Pr(X)$ by:

$$\text{SMI}[Pr(X)] = -\int Pr(X) \log_2 Pr(X) dX \tag{4.23}$$

Thus, Eq. (4.22) may be written as a difference in the SMI.

$$\Delta S_M^* = (k \ln 2)\left[\text{SMI}(Pr\left(\boldsymbol{X}^N|\boldsymbol{R}_M\right) - \text{SMI}\left(Pr\boldsymbol{X}^N\right)\right] \tag{4.24}$$

Hence, the derivative in Eq. (4.16) is written as:

$$\delta S(\sigma) = -\frac{\partial \delta G(\sigma)}{\partial T} = \Delta S_D^* - 2\Delta S_M^M$$
$$= (k \ln 2)\left[\text{SMI}\left(\Pr\left(\boldsymbol{X}^N|\boldsymbol{R}_{AB} = \sigma\right)\right) - \text{SMI}\left(\Pr\left(\boldsymbol{X}^N|\boldsymbol{R}_{AB} = \infty\right)\right)\right] \tag{4.25}$$

Thus, apart from the factor $(k \ln 2)$, the entropy change associated with the process $\infty \to \sigma$ is the difference between the conditional SMI defined on the conditional distribution of solvent configurations at initial and final states.

The interpretation of the entropy of solvation in terms of SMI is of interest in Information Theory [Ben-Naim (2017, 2022)]. In general, the SMI defined in Eq. (4.23) may be interpreted as either the uncertainty or the unlikelihood of finding the location of a particle within an interval dX at X. It is also a measure of information associated with the entire distribution $\Pr(X)$. In the case of the solvation entropy we have a *difference* in the SMI defined on the distribution of all the solvent molecules (both locations and orientations) caused by the addition of the solute M at some fixed position ($\boldsymbol{R}_M$) in the solvent.

The same interpretation applies to the solvation entropy of a "dimer." Thus, in Eq. (4.25) we have the difference in the solvation entropy of a "dimer" at two different states; at infinite distance (∞), and at some close separation σ. In this case, the interpretation of the quantity δS may be interpreted as the change in the uncertainty (or unlikelihood, or a measure of information) associated with the distribution of the configurations of all solvent molecules when we bring the "dimer" from infinite separation (∞) to some close distance σ.

Note carefully that the latter interpretation of the temperature dependence of the SII is different from the one we have discussed in connection with Eq. (4.17) in terms of change in the total average interaction energy among all the solvent molecules.

4.4 Simulated Results on $H\phi O$ Interactions and Forces

Results obtained from simulations may be viewed as an intermediary between experimental and theoretical. Simulations are some kind of "experiments" that are carried out on computers. However, unlike real experiments with real molecules, simulations are carried out with experiments on a model of the molecules involved and a model (or a theoretical derived) of the interaction between pairs of molecules. Therefore, when examining the results obtained by simulations one should remember that the results pertain to the model-system rather than the real system. In addition, there is some uncertainty about the statistical averages one calculates which is akin to the uncertainty in the accuracy of the experiments.

Specifically, when dealing with water and aqueous solutions all our models of water molecules and their interactions should be viewed at best as models that exhibit similar behavior to liquid water, not the real liquid.

Simulation of the SII between $H\phi O$ groups began soon after the $H\phi O$-bond or the $H\phi O$-effect was *defined* in terms of the indirect work (either Helmholtz or Gibbs energy) required to bring two non-polar molecules from infinite separation to contact distance σ. In fact, the first simulation results were published soon after the "Theory of $H\phi O$ effect" was published in 1977 by Pratt and Chandler.

Pangali et al. [13, 14], did a Monte Carlo simulation using two Lennard Jones particles (with diameter 4.12 Å) in "liquid water" consisting of ST2 water molecules. (The ST2 model was a modified 4-point change model known as BNS model published earlier by Ben-Naim and Stillinger. See Ben-Naim [6]).

As expected they obtained the first minimum of the PMF at about 4.2 Å, and the second minimum at about 7 Å. This distance is roughly the solvent-separated distance between the two solute molecules see Sect. 4.3 and Fig. 4.10. The results of Pangali et al. [13, 14] were in good agreement with those of Pratt and Chandler [7]. See full circles in Fig. 4.11.

In Ravishanker et al. [15], also did the Monte-Carlo simulation for two methane-like particles in water. They obtained a first minimum about 3.8 Å, and a second minimum at about 6 Å. Note that the second minimum is lower than the first one, and the distance corresponds roughly to the water-bridged configuration, see Fig. 4.10b and open circles in Fig. 4.11.

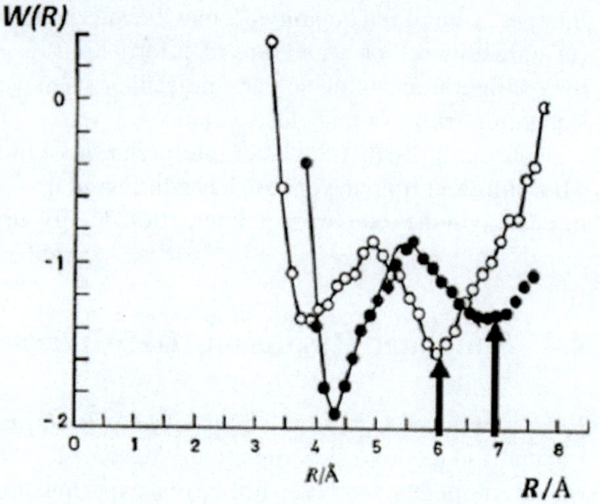

Fig. 4.11 The potential of mean force for methane in water. Based on data from Pangaly et al., Full circles, and from Ravishanker et al., open circles

New and Berne [16], did a molecular dynamic simulation of the PMF of methane, but using a different model for water. They got the first minimum at 3.8 Å, and the second at 7 Å.

More recently, Durell and Ben-Naim [17], reported simulated results on the temperature dependence of both the PMF and the force and PMF between $H\phi O$ molecules. We shall present some of these results in the next chapter.

4.5 Conditional Solvation and Conditional $H\phi O$ Interactions

We recall that the main argument Kauzmann espoused in promoting the idea of the $H\phi O$ bond was the large negative Gibbs energy change for the process of transferring a non-polar solute from water into an organic solvent. Kauzmann, apparently was not aware of the fact that there might be quite a large difference between the solvation Gibbs energy of a *solute molecule* such as methane, and solvation Gibbs energy of a group such a methyl group attached to a protein, Fig. 4.12.

Perhaps the belief that the $H\phi O$ interaction between two molecules is similar to the $H\phi O$ interaction between two groups follows from the fact that the (*direct*) interaction between methane and water or between methane and methane molecules is approximately the same as the interaction between methyl group and a water molecule or methyl and methyl group. It turns out that these two processes involve very different Gibbs energy changes. The reason is the presence of the backbone, which can have two major effects on the solvation Gibbs energy. First, in Fig. 4.12b, the water solvates the methane molecule from all directions, whereas in Fig. 4.12a the methyl group is solvated from only one side of the methyl group. Second, the

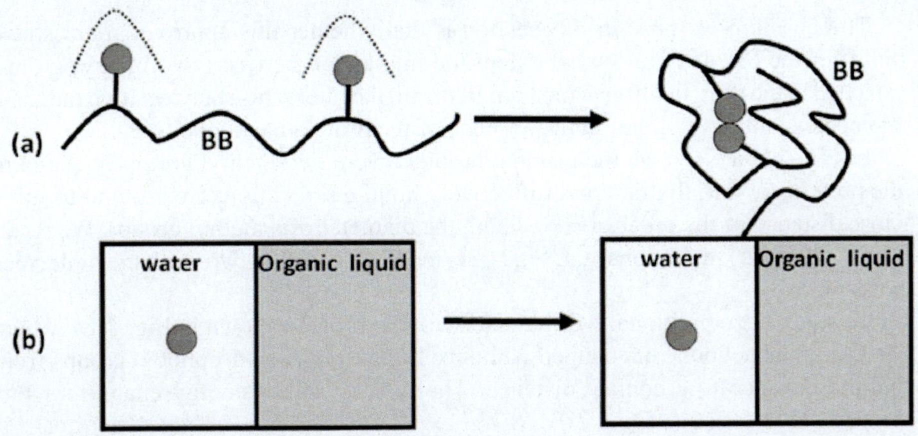

Fig. 4.12 **a** Transferring two hydrophobic groups (full circles), from being solvated by water (dashed curve), into the interior of the protein. **b** Transferring a hydrophobic solute from water into an organic liquid

protein backbone is polar, and this polarity may affect the orientation of the water molecules around the methyl group, hence also affect the solvation Gibbs energy of the methyl group. In spite of these differences, people still use Kauzmann's model as evidence for the claim that hydrophobic effect is the most important factor in protein folding.

Similarly, in my definition of the $H\phi O$ interaction I "extracted" the process of interaction of two $H\phi O$ groups attached to the protein backbone and studied the interaction between two methane molecules in water in the absence of the solvent, Fig. 4.13. Again, one can safely assume that the *direct* interaction between two methane molecules is approximately the same as the interactions between two methyl groups which are attached to the protein Fig 4.13.

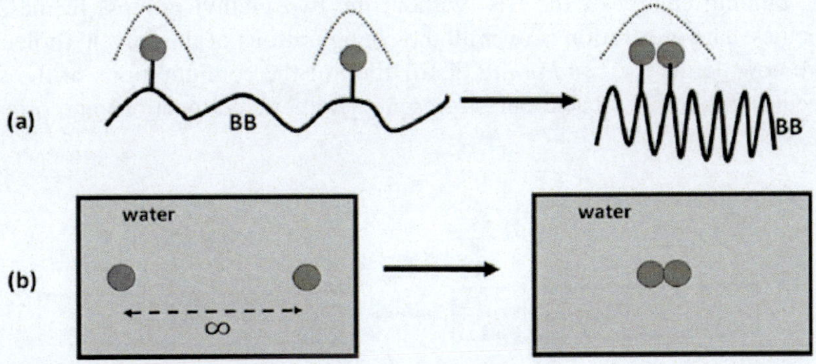

Fig. 4.13 **a** Hydrophobic interaction between two hydrophobic groups (full circles). **b** Hydrophobic interaction between two hydrophobic molecules

The question we pose in this section is that whether this approximation is also valid for the SII. If not, to what extent the interaction between two methyl groups attached to the protein differs considerably from the interaction between two methane molecules in the water, but in the absence of the protein backbone (BB).

In Fig. 4.13a we show the original problem were two methyl groups attached to the protein are initially far apart (effectively infinite separation) are brought to some close distance of the order of σ (σ being the diameter of a methyl group). We recall that in all of the simulations, e.g. Fig. 4.11, the PMF between two methane molecules was studied.

To study the conditional SII, we start with the process shown in Fig. 4.14, where we keep the backbone unchanged while we bring the two hydrophobic groups from infinite distance to a contact distance. The indirect Gibbs energy change for this process, in water at constant T, V, N is:

$$\delta A(\infty \to \sigma|BB) = -kT \ln \frac{Q(\sigma)}{Q(\infty)} \tag{4.26}$$

where $Q(\sigma)$ and $Q(\infty)$ are the canonical PF of the system T, V, N having the BB and two methyl groups at distance σ and ∞, respectively.

We assume, as usual, that all the internal PFs of the molecules in the system do not change in this process. Therefore, the ratio of the canonical PFs may be written as the ratio of the two configurational PFs in the initial and final states:

$$\frac{Q(\sigma)}{Q(\infty)} = \frac{\int \exp[-\beta U_N - \beta B_D(\sigma) - \beta B_{BB}]d\boldsymbol{X}^N}{\int \exp[-\beta U_N - \beta B_D(\infty) - \beta B_{BB}]d\boldsymbol{X}^N} \tag{4.27}$$

In Eq. (4.27), U_N is the total interaction energy among all the N water molecules at a specific configuration $\boldsymbol{X}^N$ (which is omitted for simplification). $B_D(\sigma)$ is the binding energy of the "dimer" (i.e. the two methyl groups) at a distance σ, and similarly $B_D(\infty)$ is the binding energy of the "dimer" at infinite separation. B_{BB} is the binding energy of the BB (without the two methyl groups) to the solvent molecules. The integration is over all the configurations of the solvent molecules.

We now define the *conditional* distribution of the configurations of the solvent molecules given that a backbone is present at some fixed locations and orientation in the system by:

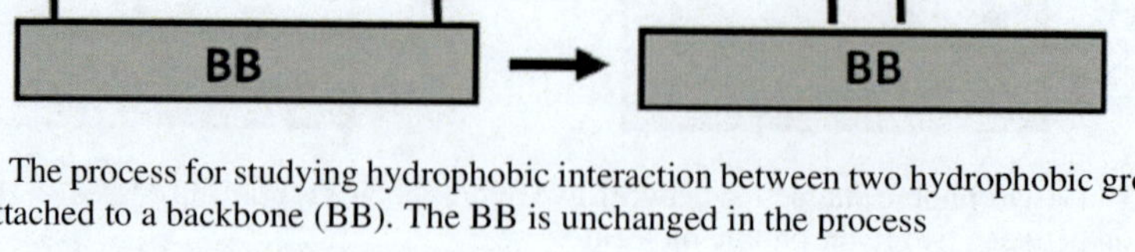

Fig. 4.14 The process for studying hydrophobic interaction between two hydrophobic groups (full circles), attached to a backbone (BB). The BB is unchanged in the process

$$Pr(X^N|BB) = \frac{\exp[-\beta U_N - \beta B_{BB}]}{\int \exp[-\beta U_N - \beta B_{BB}]dX^N} \tag{4.28}$$

with this distribution, we can rewrite Eq. (4.27) as:

$$\frac{Q(\sigma)}{Q(\infty)} = \frac{\int \exp[-\beta(U_N - B_D(\sigma) + B_{BB})]dX^N}{\int \exp[-\beta(U_N - B_{BB})]dX^N} \times \frac{\int \exp[-\beta(U_N + B_{BB})]dX^N}{\int \exp[-\beta(U_N + B_D(\infty) + B_{BB})]dX^N} \tag{4.29}$$

We now recognize that these two quotients are related to the ratio of two solvation Gibbs energies. Therefore, we rewrite Eq. (4.26) using Eq. (4.29) as:

$$\begin{aligned}\delta A(\infty \to \sigma|BB) &= -kT \ln \frac{Q(\sigma)}{Q(\infty)} \\ &= -kT \ln\left[\frac{\langle \exp[-\beta B_D(\sigma)]\rangle_{BB}}{\langle \exp[-\beta B_D(\infty)]\rangle_{BB}}\right] \\ &= \Delta A^*(\sigma|BB) - \Delta A^*(\infty|BB)\end{aligned} \tag{4.30}$$

where $\Delta A^*(\sigma|BB)$ is the conditional solvation Helmholtz energy of the system in the final state in Fig. 4.14, and $\Delta A^*(\infty|BB)$ is the conditional solvation Helmholtz energy at the initial state. The notation $\langle\rangle_{BB}$ means average over all the configurations of the solvent molecules with the conditional distribution function defined in Eq. (4.28).

The quantity $\delta A(\infty \to \sigma|BB)$, or the corresponding Gibbs energy change $\delta G(\infty \to \sigma|BB)$ is the more realistic measure of the SII between two methyl groups attached to the protein. Originally, this was referred to as *intramolecular* $H\phi O$ interaction, Ben-Naim and Wilf [18]. However, the term *conditional* SII is more appropriate.

Equation (4.30) is the correct contribution of one pair of methyl groups to the overall effect of the solvent on the process of protein folding. In practice, we cannot use this equation because we do not have the solvation Gibbs (or Helmholtz) energies of solvation on the right-hand side of the equation. Therefore, we turn to study model compounds for which such information is available.

Consider the process shown in Fig. 4.15a. We have a backbone (the hexagon) on which two methyl groups are attached. We assume that the bond energy of the methyl group to the BB is independent of the location at which the methyl group is attached (we refer here only to the $c - c$ chemical bond between the carbon of the methyl group and the carbon of the protein. We do not include direct interaction energy between the two methyl groups).

The change in Gibbs energy for the process shown in Fig. 4.15a at constant (P, T, N) can be written as:

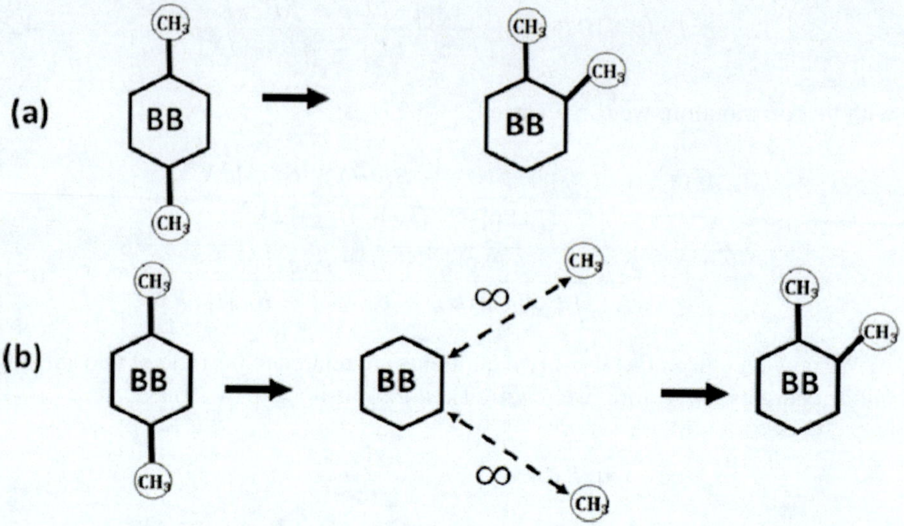

Fig. 4.15 **a** The transfer of a methyl group from position "4" to position "2" on the benzene ring. **b** The same process as in **a** but we first remove the two methyl groups to infinite separation, then bring them back but to the "1, 2" positions

$$\Delta G(a) = U(a) + \delta G(a) \tag{4.31}$$

The indirect part of this interaction is related to the difference in solvation Gibbs energies as follows:

$$\delta G(a) = \Delta G^*(\sigma) - \Delta G^*(\infty) \tag{4.32}$$

This equation can be established either by writing the PF of the system in the initial (∞) and the final states, or simply by carrying out the process "a" in Fig. 4.15a, once in the liquid, and once in a vacuum. The total Gibbs energy change for the process in Fig. 4.15a is:

$$\begin{aligned}\Delta G(''\infty'' \to \sigma) &= U(a) + \delta G(a)\\ &= U(a) - \Delta G(''\infty'') + \Delta G^*(\sigma)\end{aligned} \tag{4.33}$$

Here $-\Delta G(''\infty'')$ and $\Delta G^*(\sigma)$ are the solvation Gibbs eneries of the left and the right compounds in Fig. 4.15a. In the process (a) in the liquid we have *direct* and the indirect contributions. The same process (a) can be achieved by first desolvation, the molecule at state "∞," then performing (a) in vacuum, and finally solvating the molecule in state "σ." (The symbol "∞" means here that the two methyl groups are at a very large distance, such that their solvation spheres do not overlap). This assumption will be examined below).

Clearly, if BB is the entire protein we clearly cannot measure the two quantities on the right-hand side of Eq. (4.32). However, if we can find a BB which is small enough, we might be able to measure the solvation Gibbs energies of the two molecules in Eq. (4.32). It is easy to show that $\delta G(a)$ in Eq. (4.32) is also equal to:

$$\begin{aligned}\delta G(a) &= \Delta G^*(\sigma) - \Delta G^*(\infty) \\ &= \Delta G^*(BB) + \Delta G^*(\sigma|BB) \\ &\quad - \Delta G^*(BB) - \Delta G(\infty|BB) \\ &= \Delta G^*(\sigma|BB) - \Delta G^*(\infty|BB)\end{aligned} \tag{4.34}$$

This means that in order to measure the solvation Gibbs energy of either the molecules on the left or on the right on Fig. 4.15a, we first solvate the BB, then we conditionally solvate the two methyl groups. Since the solvation Gibbs energy of BB cancels out, we get the required quantity. Compare Eq. (4.34) with (4.30). We see that it is exactly the same equation except that on the right-hand side of (4.34) we have measurable quantities.

Before we present some numerical data, we should explain the notation "∞" for the distance between the two methyl groups in the initial state in Fig. 4.15a. We explain this notation by using a specific example of a backbone, the benzene ring. In this case, the process in Fig. 4.16a involves the transfer of a methyl group from position "4" in the dimethyl benzene into position "2," where position "1" is already occupied by a methyl group. The assumption is made that in the 1,4-dimethyl benzene the distance between the two methyl groups is large enough so that the two methyl groups are independently solvated. This is tantamount to saying that the distance between the two methyl groups is practically "infinity." To check the validity of this assumption we perform another "reaction" as shown in Fig. 4.16b. We start with two methyl-benzene molecules and transfer one methyl group from one molecule (at infinite separation) to the second molecule to form 1,2-dimethyl benzene. The solvent-induced change in Gibbs energy for this case is:

$$\delta G(b) = \Delta G^*(\sigma) + \Delta G^*(BB) - 2\Delta G^*(1) \tag{4.35}$$

where $\Delta G^*(1)$ is the solvation Gibbs energy of the methyl-benzene (toluene), and $\Delta G^*(BB)$ is the solvation Gibbs energy of the BB, which in this case is the benzene ring (we ignore in all our discussions the contribution of the hydrogen atom to the solvation Gibbs energy of the benzene molecule). Similarly, we do the same for the 1,4 isomer, Fig. 4.16c, for which we write:

$$\delta G(c) = \Delta G^*(\infty) + \Delta G^*(BB) - 2\Delta G^*(1) \tag{4.36}$$

By subtracting Eq. (4.36) from (4.35), we get:

$$\begin{aligned}\delta G(c) - \delta G(b) &= \Delta G^*(\infty) - \Delta G^*(\sigma) \\ &= \Delta G^*(\infty|BB) - \Delta G^*(\sigma|BB)\end{aligned} \tag{4.37}$$

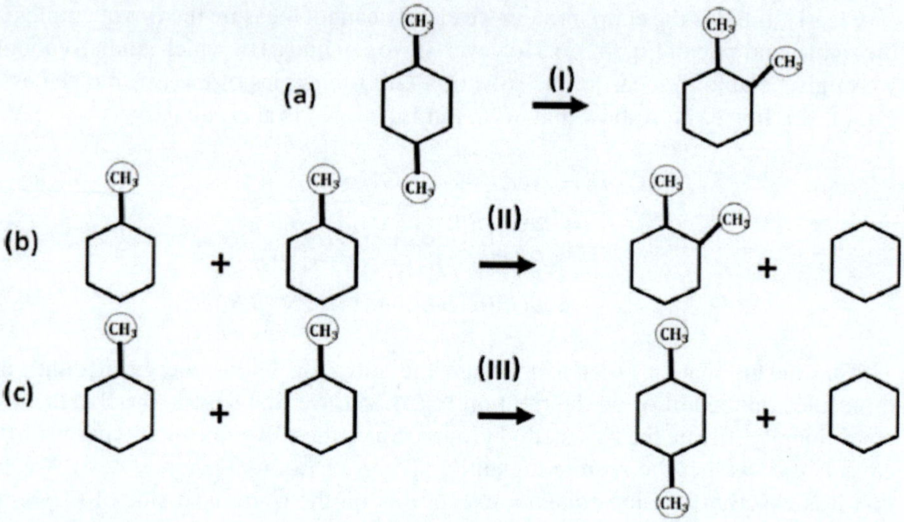

Fig. 4.16 The two processes of transfer **b** and **c** used to justify the assumption that the distance between the two methyl groups in the 1,4 isomer **a** is nearly infinity, see Sect. 4.5

Note that here we use the infinity sign because at the initial state the distance between the two methyl groups is indeed such that the two methyl groups are independently solvated. If the values of $\delta G(c)$ in the last equation will equal the value of $\delta G(b)$, then we can conclude that our assumption about the distance (denoted "∞") between the two methyl groups in the 1,4 isomer is valid. Actually, the experimental data shows that the values $\delta G(c)$ are nearly equal to those $\delta G(b)$. For details see Ben-Naim [9].

Table 4.2 shows the values of $\delta G(a)$ in H_2O, D_2O, and in n-hexane at two temperatures.

From Table 4.2 we see that the SII between two methyl groups in the presence of the benzene ring is about 0.3 kcal/mol in H_2O and slightly smaller (in absolute magnitude) in D_2O. This is an important result. It shows that the $H\phi O$ interaction is not necessarily stronger when the "peculiarities" of water are larger. This is similar to the comment we made in connection with the conclusion of Pratt and Chandler, about the temperature dependence of the $H\phi O$ interaction, in Sect. 4.3. D_2O is considered more "structured" liquid than water, yet the $H\phi O$ interaction are slightly weaker. Thus "more structured" does not necessarily implies "stronger $H\phi O$ interaction."

Table 4.2 Values of $\delta G(a)$ (in cal/mol), for the transfer of methyl group from position "4" to position "2," in H_2O, D_2O, and n-hexane at two temperatures

t °C	H_2O	D_2O	n-hexane
10	−238		70
20	−300	−280	−28

Table 4.3 Values of $\delta G(a)$ (in cal/mol) for the transfer of ethyl group from position "4" to position "2," in H_2O, D_2O, and n-hexane at two temperatures

t °C	H_2O	D_2O	n-hexane
10	−603		+360
20	−616	−430	+445

Note also that the temperature dependence of the $H\phi O$ interaction between 10 °C and 20 °C is the same as we observed for two methane molecules.

The value $\delta G(a)$ for the same process in n-hexane is much smaller, and in fact, these values are within the experimental error.

From Table 4.3 we see that the values of $\delta G(a)$ for transferring the ethyl group are almost double the value for the methyl group. However, the temperature dependence and the difference between H_2O and D_2O are similar in the two cases. Note also that $\delta G(a)$ in n-hexane in Table 4.3 are large and *positive*, not negative as in either H_2O or D_2O.

In conclusion, the values of δG presented in Tables 4.2 and 4.3 are more realistic than the values for $\delta G(\sigma_1)$ reported in Sect. 4.4, for two reasons. First, because the distance between the two methyl groups is more realistic than the distance between two methyl groups in ethane. Second, we have here the presence of a BB. Although, this is not exactly the BB of a protein, it still gives us an idea how the presence of the BB might affect the SII.

4.6 Hydrophobic Interactions Among Many Non-polar Molecules

As we have noted in Chap. 1, Kauzmann's model for the $H\phi O$-bond was initially inspired by the fact that most of the $H\phi O$ side chains of the amino acids in globular proteins are found in the interior of the protein. Clearly, these groups are removed from the water environment, and the tendency to avoid water was termed the hydrophobic effect. Similarly, the tendency of the non-polar groups to "adhere to each other" was termed (inappropriately) as hydrophobic bonding.

In this section we discuss the question of the tendency of a large number of non-polar molecules to form aggregates in aqueous solution. We consider the process of n non-polar solute molecules forming a spherical aggregate in water, Fig. 4.17.

Clearly, when n is very large, most of the solute particles are desolvated by water and are found in the interior of the aggregate. Those particles which are on the surface of the aggregates are only partially exposed to the solvent.

For small n we already reported in section ... on the SII between four methane molecules brought to the configuration to the peripheral methyl groups in neopentane. In calculating δG for that process, we used the same approximation as the one we

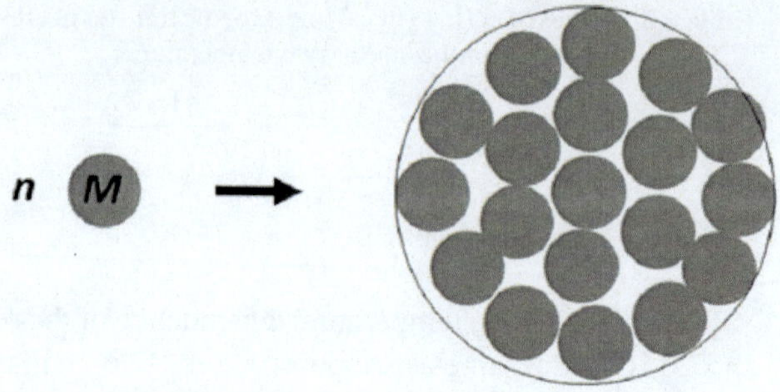

Fig. 4.17 Aggregation of n monomers (M) forming an approximate spherical aggregate

used in Sect. 4.1 to calculate the pairwise interaction between two methane molecules at very short distance.

In the following we describe some results obtained by using the Scaled-Particle Theory (SPT).

We denote the aggregation reaction in which n simple solute molecules form a spherical aggregate of diameter σ_A as:

$$nM \rightarrow Agg. \tag{4.38}$$

The corresponding Gibbs energy change for bringing the n monomer M from fixed positions but at infinite separation from each other to the final configuration of the aggregate ($Agg.$) at some fixed position in the solvent as:

$$\Delta G(\infty \rightarrow Agg.) = U_n + \delta G(Agg.) \tag{4.39}$$

And for the solvent-induced part of ΔG we can write:

$$\delta G(Agg.) = \Delta G^*_{Agg.} - n\Delta G^*_M \tag{4.40}$$

ΔG^*_M and $\Delta G^*_{Agg.}$ are the solvation Gibbs energies of the monomer (M) and the aggregate (Agg), respectively. Both of these quantities may be computed by the SPT. However, a better approximation would be to take the experimental values of ΔG^*_M for a non-polar solute such as argon or methane, and to use the SPT to calculate the cavity-work for the aggregate. The idea is that for very large spherical aggregates the main contribution to ΔG^*_{Agg} would come from the solvation Gibbs energy of a hard-sphere having approximately the same diameter as the aggregate, i.e. we assume that:

$$\Delta G^*_{Agg.} \approx \Delta G^*_{HS}(\sigma_A) = \Delta G^*_{cav}(R_{cav}) \tag{4.41}$$

σ_A is the diameter of the aggregate (or the equivalent hard-sphere), and R_{cav} is the radius of the cavity which is equivalent to a HS of diameter σ_A, i.e.

$$R_{cav} = \frac{\sigma_A + \sigma_S}{2} \tag{4.42}$$

Thus, the SII among the n solute particles forming a spherical aggregate is approximated by:

$$\delta G(Agg.) \approx \Delta G^*_{cav}(R_{cav}) - n\Delta G^*_M \tag{4.43}$$

where the quantity ΔG^*_M is taken from experimental sources and $\Delta G^*_{cav}(R_{cav})$ is calculated from SPT.

We use our solute as methane, with a diameter of $\sigma_A = 3.82$ Å. In order to use the SPT, we need to estimate the radius of the cavity formed by the aggregate of n monomers. If n is large enough and all the monomers are packed compactly in the aggregate of diameter σ_A, then the ratio of the volume of the n monomers and the volume of the hard-sphere of diameter σ_A is:

$$n \times \frac{\pi\sigma_M^3/6}{\pi\sigma_A^3/6} = 0.7405 \tag{4.44}$$

from which we get σ_A as:

$$\sigma_A \approx \left(\frac{n\sigma_M^3}{0.7405}\right)^{1/3} \tag{4.45}$$

Once we have the diameter of the aggregate we can calculate the radius of the corresponding cavity, and use the SPT to estimate $\Delta G^*_{cav}(R_{cav})$.

In Fig. 4.18 we present some values of $\delta G(Agg.)$ as a function of n for different solvents. We choose the diameter $\sigma_A = 3.82$ Å for methane and $\sigma_W = 2.90$ Å for water, $\sigma_{met} = 3.7$ Å for methanol $\sigma_{Et} = 4.34$ Å for ethanol and $\sigma_c = 5.63$ Å for cyclodrexane.

From Fig. 4.18 it is clear that the values of $\delta G(Agg.)$ in water and heavy water are negative and its dependence on n is much greater than in all the other liquids. Clearly, the large negative values of $\delta G(Agg.)$ reflects the tendency of the non-polar molecule to avoid the water (i.e. desolvation of methane is large and negative). This tendency becomes larger the larger the aggregate. The reason is simple, in large spherical aggregates the ratio of the molecules in the interior to those at the surface of the aggregate is larger the larger n.

We note also that $\delta H(Agg.)$ and $\delta S(Agg.)$ are quite small in all liquids. However, for water and heavy water both the enthalpy and the entropy change are large and positive. This is, again consistent with the experimental fact that the solvation enthalpy and entropy of methane in water are large and negative in water (due to increase in the total number of HBs in the water—a phenomenon described qualitatively as in increase in the structure of water. See Ben-Naim [6]. The process of aggregation (of large n) may be viewed as desolvation as methane molecules, hence, we expect a large positive value for both the entropy and the enthalpy changes when the process occurs in water.

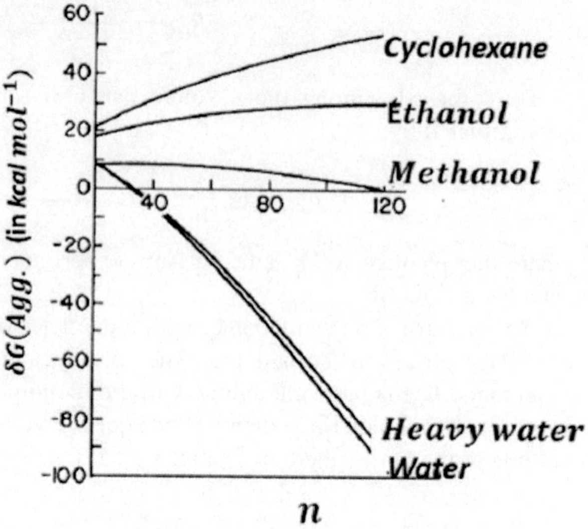

Fig. 4.18 Values of $\delta G(Agg.)$ as a function of n, the size of the aggregates, in different solvents

Thus, we conclude that in spite of the extreme approximation of the SPT, all the results we obtained agree with what we already know from other sources.

References

1. Schrier EE, Pottle M, Scheraga HA (1964) JACS 86:3444
2. Kozak JJ, Knight WS, Kauzmann W (1968) J Chem Phys 48:675
3. Ben-Naim A (1980) Hydrophobic interactions. Plenum Press, New York
4. Ben-Naim A (1987) Solvation thermodynamics. Plenum Press, New York
5. Ben-Naim A (1974) Water and aqueous solutions, introduction to a molecular theory. Plenum Press, New York
6. Ben-Naim A (2009) Molecular theory of water and aqueous solutions, part I: understanding water. World Scientific, Singapore
7. Pratt LR, Chandler D (1977) J Chem Phys 67:3683
8. Reiss H (1966) Adv Chem Phys 9:1
9. Ben-Naim A (1992) Statistical thermodynamics for chemists and biochemists. Plenum Press, New York
10. Pierotti RA (1963) J Phys Chem 67:1840
11. Pierotti RA (1965) J Phys Chem 69:281
12. Ben-Naim A (2011) Molecular theory of water and aqueous solutions, part II: the role if water in protein folding, self-assembly and molecular recognition. World Scientific, Singapore
13. Pangali C, Rao M, Berne BJ (1979) J Chem Phys 71:2975
14. Pangali C, Rao M, Berne BJ (1979) J Chem Phys 71:2982
15. Ravishanker G, Mezei M, Beveridge DL (1982) Faraday Simp Chem Sos 17:79
16. New MH, Berne BJ (1995) J Am Chem Soc 117:7172–7179
17. Durell SR, Ben-Naim A (2017) Hydrophobic-hydrophilic forces in protein folding. Biopolymers 10:1–14
18. Ben-Naim A, Wilf J (1979) A direct measurement of intramolecular hydrophobic interactions. J Chem Phys 70:771–777

Chapter 5
Hydrophilic Interactions and Forces

The term "hydrophilic" ($H\phi I$) was used long before the term "hydrophobic" was introduced in connection with the problem of protein folding. It should be noted that "water fearing" is not exactly the opposite of "water-loving." While the term $H\phi O$-bond, or later the term $H\phi O$ interactions were extensively used in connection with the protein folding problem, the term "hydrophilic interaction" was not even mentioned in this context until the late 1980s. It was only after a complete and thorough inventory of all possible solvent-induced effects involved in protein stability, became available, that the $H\phi I$ interactions appeared in the literature. As I have explained in the preface and in Chap. 1, I spent over 15 years studying $H\phi O$ interactions, in general, and in connection with the protein folding problem, in particular. These works were published in my book on *Hydrophobic Interactions* in 1980. A few years later, after developing an "inventory" of all possible SII, and after seeing some relevant experimental data, I was immensely surprised to find that $H\phi I$ interactions are far stronger than the corresponding $H\phi O$ effects, in all biochemical processes in which $H\phi O$ effects were previously involved. It is now clear that the evidence favoring $H\phi I$ over $H\phi O$ effects are overwhelming not only in explaining and determining the stability and specificity of the structure of proteins, but also in explaining the phenomena of self-assembly of proteins as well as molecular recognition.

In this chapter, we survey the evidence regarding various $H\phi I$ effects. We start with some theoretical estimates, then survey the experimental evidence and finally report on some simulated results. All of this evidence is very approximate. The theoretical estimates are based on very crude assumptions. The experimental results, even when they are quite accurate, are relevant to some specific configurations, which might not be the ones that are of interest in connection with protein stability. Finally, the simulated results depend on the model one uses for water-water pair potential. However, in spite of all the shortcomings, as pointed out above, the results reported in this chapter clearly indicate that the values of the $H\phi I$ effects are much larger than the corresponding values of the $H\phi O$ effects.

A. Ben-Naim, *Solvent-Induced Interactions and Forces in Protein Folding*, Physical Chemistry in Action, https://doi.org/10.1007/978-3-031-31871-9_5

5.1 Theoretical Evidence on $H\phi I$ Effects

In this chapter, we shall focus mainly on pairwise $H\phi I$ interactions. There are essentially two such pairwise $H\phi I$ interactions:

The first, is the process of bringing two $H\phi I$ groups (these could either be hydroxyl, OH, carbonyl, C=O, or amine, either NH or NH_2 in proteins), from infinite separation to the final configuration at which the two groups from *direct* HB. In most of the simulated calculations, we use the OH and the lone-pair directions of a water molecule instead of a donor and an acceptor hydrophilic groups. The process is shown in Fig. 1.15 and the final configuration in Fig. 5.1. The chemical reaction is:

$$R - \mathrm{OH} + \mathrm{O} = \mathrm{C} - R \rightarrow R - \mathrm{OH} \cdots \mathrm{O} = \mathrm{C} - R \quad (5.1)$$

Here a molecule having a hydroxyl group (OH), forms a HB with a molecule having a carbonyl group (C=O). As we have discussed in Chap. 1, this kind of *direct* formation of HB was deemed, by Kauzmann and others, to be insignificant to protein stability. In fact, this (erroneous) conclusion was the basis that led Kauzmann to suggest the idea of the $H\phi O$-bond, which idea that reigned supreme in protein chemistry literature for almost half a century.

The second $H\phi I$ effect was discovered only in the late 1980s, and in my opinion is the most important in protein folding (as well as in self-assembly of proteins, molecular recognition, and solubility of globular proteins, see Chaps. 1 and 6). The process is shown in Fig. 1.16 and the final configuration is shown in Fig. 5.2. Two $H\phi I$ groups are brought from infinite separation to a distance and to a specific orientation such that they can be bridged by water molecules Fig. 5.2.

We shall discuss separately the theoretical evidence for these two pairwise $H\phi I$ effects.

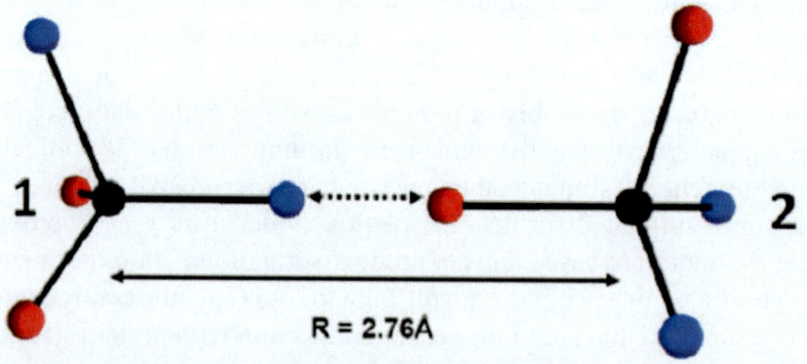

Fig. 5.1 The final configuration of a direct HB, between an OH (blue) of one water molecule and the lone-pair direction (red) of a second water molecule. Hypothetical process in which a permanent HB–bridge is formed

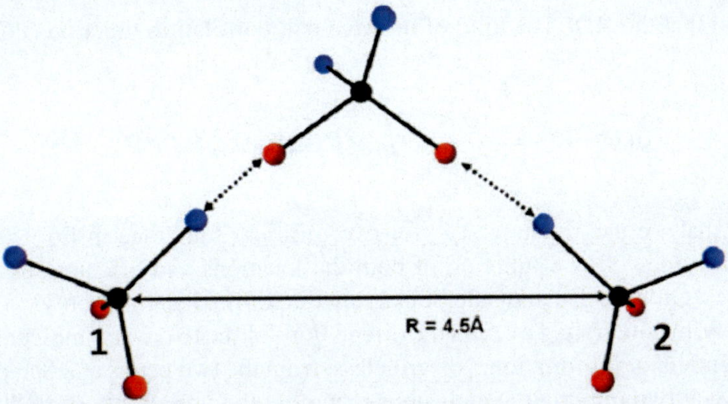

Fig. 5.2 The final configuration in which HB–bridge is formed between two OH groups (blue) of two water molecules

5.1.1 Direct Formation of HB

Figure 5.3 shows the radial distribution function of liquid water at 25 °C. There are two pronounced peaks at about 2.8 Å and 4.5 Å. Corresponding to the reduced distance R* = 1 and 1.6 respectively. See the dashed arrows in Fig. 5.3. In a liquid consisting of simple spherical particles (e, g. argon), we expect to find the first two peaks at about R* = 1 and 2 respectively.

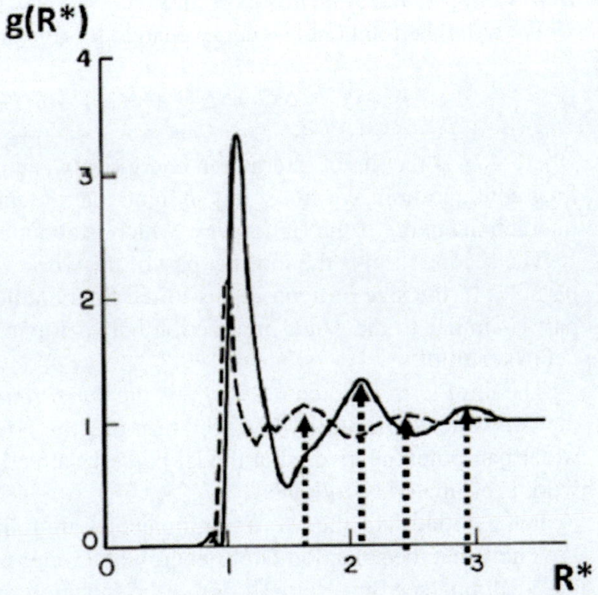

Fig. 5.3 The radial distribution function of water (dashed curved) at 4 °C, and of liquid argon (full curve) at 84 K. R* is the reduced distance R* = R/σ, where σ Is the diameter of the molecule. The dashed arrows show the locations of the first and the second peaks in the RDF

Recall that the RDF is a kind of *average* pair correlation function (PCF) defined by:

$$g(\boldsymbol{R}_1, \boldsymbol{R}_2) = \frac{1}{(8\pi^2)^2} \int g(\boldsymbol{R}_1, \boldsymbol{R}_2, \boldsymbol{\Omega}_1, \boldsymbol{\Omega}_2) d\boldsymbol{\Omega}_1, d\boldsymbol{\Omega}_2 \tag{5.2}$$

Note that we use the letter "g" for two different functions in Eq. (5.2). On the right-hand side "g" is a function of both the locations and orientations of the two molecules. On the left-hand side, it is a function of the locations only.

Clearly, by integrating over all the orientations of the two water molecules we lose a great amount of information. Nevertheless, from the two peaks at 2.8 and 4.5 Å, we can arrive at two important conclusions regarding the probability of finding the two water molecules at these two distances. The RDF does not provide any information on the relative orientations of the two water molecules at these probable distances. In other words, we cannot "de-average" the RDF to obtain the PCF. Instead, we shall try to estimate the Gibbs energy change for the specific process of bringing the two water molecules from infinite separation to the final distance R. Then estimate the value of the PCF at this configuration.

We view each water molecule as consisting of a Lennard–Jones sphere and four unit-vectors (or "arms") along which the molecules can form HB. We distinguish between two arms along which are the OH direction of the water molecules along which the molecule can form a HB as a "donor, "i.e. donor of the OH to the HB. The other two directions are along which the water molecule forms a HB as an acceptor. These arms will be referred to as the lone-pair (LP) direction. (For more details, see Ben-Naim [1], and Appendices A and B).

We write the total Gibbs energy change for this process (Fig. 1.15) as:

$$\Delta G = \Delta U + \delta G_{LJ} + \delta G_{HB|LJ} \tag{5.3}$$

where ΔU is the direct interaction energy between the two water molecules at the final configuration. We can safely assume that the most important part of the direct interaction energy is the HB energy, which is about $\varepsilon_{HB} \approx -6.5\,\text{kcal/mol}$.

The second term is the indirect part of the work associated with bringing two LJ particles of the size of neon atoms to the final configuration. We estimate that this part is similar to the $H\phi O$ interaction between two neon particles which is about – 0.6 kcal/mol.

The third term denoted $\delta G_{HB|LJ}$ is the *conditional* indirect interaction between the two water molecules when we "turn on" the HB interaction part of the water-water pair potential given that the LJ part was already "turned on." This part of the work is estimated as follows.

In the initial state, the two water molecules are fully solvated by water molecules. We know that the solvation Gibbs energy of a water molecule due to HBing is about – 9 kcal/mol (see Ben-Naim [1, 2]), or the solvation per arm is about – 2.25 kcal/mol when the two water molecules are at a fixed position and at infinite separation, the solvation Gibbs energy of the pair is about $-2 \times 9 = -18$ kcal/mol. In the final configuration, two of the arms are engaged in direct HBs. Therefore, the pair of water molecules at this configuration is solvated only by six arms. Therefore, we can estimate this part of the indirect Gibbs energy change as:

$$\delta G_{HB|LJ} \approx 6 \times (-2.25) - 8 \times (-2.25) = 4.5\,\text{kcal/mol} \tag{5.4}$$

Thus, when a HB is formed two arms lose their solvation Gibbs energy which means a change of about 4.5 kcal/mol.

Thus, the total Gibbs energy change for the process shown in Fig. 1.15, is estimated as:

$$\Delta G(direct HB) \approx \Delta U + \delta G_{LJ} + \delta G_{HB|LJ}$$

$$\approx -6.5 - 0.6 + 4.5 \approx -2.6\,\text{kcal/mol} \tag{5.5}$$

Recall that in the early literature the Gibbs energy change for this reaction was estimated to be zero (due to the faulty HB-inventory argument (see Chap. 1).

This value of ΔG (at the final configuration of Fig. 1.15) may be converted to the pair correlation function, say at 25 °C by the equation:

$$g\left(R = 2.76\,\text{Å}, \Omega_{HB}\right) = \exp[-\beta \Delta G] \approx 80 \tag{5.6}$$

This is the value of the full pair correlation function at a distance of $R = 2.76$ Å, and at the specific orientation as in Fig. 5.1.

The experimental value of the RDF of water at the first maximum is about 2. This drastic reduction from the value of about 80–2 is due to the averaging over all possible orientations of the two water molecules. Note that this average Eq. (5.2), gives equal probability to each orientation of the water molecules. It is clear that the integral in Eq. (5.2), gets considerable contribution only from a small region in the configurational space of the two water molecules. In most of the non-HBing configuration, $\delta G_{HB|LJ}$ would be zero.

We can conclude that direct HBs contribute significantly to the Gibbs energy change in the process of Fig. 1.15. We can also conclude that a similar contribution may be expedited from two $H\phi I$ groups attached to the protein.

5.1.2 *Interaction Through a Water-Bridge*

The second important $H\phi I$ interaction is shown in Fig. 5.2. This interaction had never been featured in the literature on protein chemistry. It will be shown that this kind of $H\phi I$ interaction is probably the most important ones in the stability of the

3-D structure of the protein. We present here three estimates of the change in Gibbs energy for the process shown in Fig. 1.16, which we denote by ΔG (*HB bridge*).

1. The first estimate is based on the existence of the second peak in the radial distribution function for water, Fig. 5.3. The existence of this peak, in the experimental RDF of water, indicates that two water molecules in liquid water have a relatively high probability to be found at a distance of $R = 4.5$ Å. The RDF does not tell us anything about the orientations of the two water molecules at this distance, see Fig. 5.2.

   We now do a very rough and very inaccurate estimate of the minimum value of $\Delta G(HBbridge)$ using an argument similar to that used in Sect. 5.1.1. In the previous section, we estimated the maximum value of $g(R = 2.76\,\text{Å}, \Omega_1, \Omega_2)$ to be about 80, see Eq. (5.6). In other words, we may say that starting with the full pair correlation function $g(R = 2.76\,\text{Å}, \Omega_1, \Omega_2)$, and then integrating over all possible configurations, we get the value of the radial distribution function at this distance of $R = 2.76$ Å, i.e. we write this as:

$$\frac{1}{(8\pi^2)^2}\int g(R = 2.76\,\text{Å}, \Omega_1, \Omega_2)\mathrm{d}\Omega_1, \mathrm{d}\Omega_2 = g(R = 2.76\,\text{Å}) \approx 2 \quad (5.7)$$

   Thus, by fixing the distance at $R = 2.76$ Å and averaging all orientations of the two molecules we reduce the *maximal* value of the full pair correlation function by a factor of 40.

   Recall that the integration in Eq. (5.7) is an average provided we assume that all configurations have the same probability.

   We now make the following heuristic assumption regarding the relationship between the maximal value of the full pair-correlation function $g(R = 4.5\,\text{Å}, \Omega_1, \Omega_2)$, and the maximum value of the radial distribution function $g(R = 4.5\,\text{Å})$ which is experimentally known to be:

$$\overline{g}(R = 4.5\,\text{Å}) \approx 1.2 \quad (5.8)$$

   We now write the analogue of Eq. (5.7) as:

$$\frac{1}{(8\pi^2)^2}\int g(R = 4.5\,\text{Å}, \Omega_1, \Omega_2)\mathrm{d}\Omega_1, \mathrm{d}\Omega_2 = g(R = 4.5\,\text{Å}) \approx 1.2 \quad (5.9)$$

   We now use Eq. (5.9) to estimate the maximal value of $g(R = 4.5\,\text{Å}, \Omega_{HBB})$, i.e. the value of the full pair correlation function $g(R = 4.5\,\text{Å}, \Omega_1, \Omega_2)$ at a specific configuration denoted Ω_{HBB} at which the two molecules can form a HB bridge.

   To do that we remember that the ratio between $g(R = 2.76\,\text{Å}, \Omega_{HB})$, and $g(2.76\,\text{Å})$ was about 40 (see Sect. 5.1.1). Applying the same ratio to the distance $R = 4.5$ Å, we get:

$$g(R = 4.5\,\text{Å}, \Omega_{HBB}) \approx 40 \times 1.2 \approx 48 \tag{5.10}$$

This value may be converted into PMF at $R = 4.5$ Å to obtain (at 25 °C) the value of:

$$W(R = 4.5\,\text{Å}, \Omega_{HBB}) = -kT \ln g(R = 4.5\,\text{Å}, \Omega_{HBB})$$
$$= -1.98 \times 298 \ln 48 = -2.29\,\text{kcal/mol} \tag{5.11}$$

This is a very qualitative argument, yet the result obtained in Eq. (5.11) is quite close to the other estimates as well as with results obtained from simulations and experimental data.

2. The second estimate is based on "counting" the change in the solvated arms. It is similar to the calculation made in Sect. 5.1.1 for the direct HB.

   Recall that the configuration shown in Fig. 5.1, is one where we lost the solvation Gibbs energies of *two arms,* and gained one direct HB energy.

   We now do the following *hypothetical reaction.* We start with three water molecules, all at fixed configurations as shown in Fig. 5.2. We bring two water molecules from a fixed configuration, but at infinite separation to the final configuration as in Fig. 5.4, but now we *assume* that a *permanent* HB-bridge is formed between the two water molecules. The final configuration in Fig. 5.4 is the same as in Fig. 5.2.

   As we have already noted in Chap. 1, when we refer to a HB bridge we do not mean two permanent HBs, but a statistical bridge. This means that when the two water molecules are at configuration shown in Fig. 5.2, they *can* form a HB bridge, but this bridge is not permanent, but only with high probability.

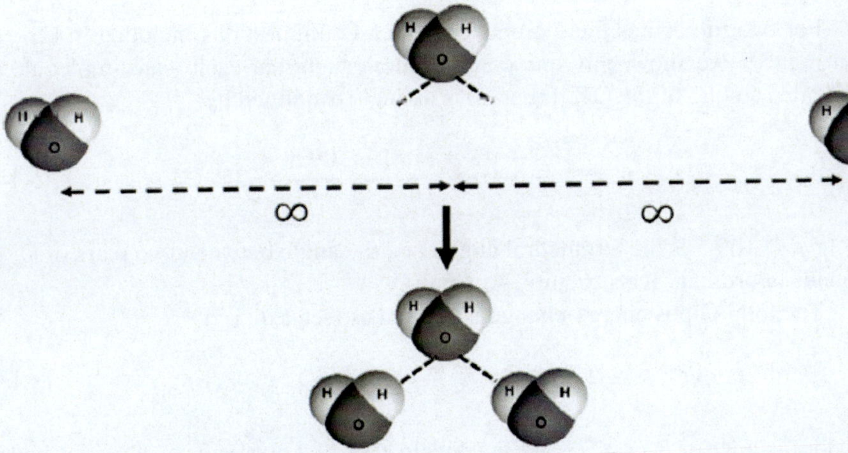

Fig. 5.4 Hypothetical process in which three water molecules, initially at infinite separation, form a permanent HB–bridge (HBB)

Thus, using the unrealistic assumption that there is a permanent HB-bridge we can "count" the loss of four solvation of the four arms, and the gain of two HB energies. Therefore, in this hypothetical reaction we can estimate:

$$\begin{aligned} &\Delta G(permanent\ HB - bridge) \approx \\ &- 4 \times (2.25) + 2 \times (-6.5) = -4\,\text{kcal/mol} \end{aligned} \tag{5.12}$$

Note again that the process described in Fig. 1.16 (with a permanent HB bridge), is a hypothetical one. When we bring two water molecules from infinite separation to the final configuration such as in Fig. 5.2, we do not expect to find there a water molecule ready to form a permanent HB-bridge. Therefore, we do not expect to "gain" two HB-energies in this process. Therefore, the value we obtain in Eq. (5.12) is deemed to be unrealistic. In the next case, we turn to a more realistic process to estimate the Gibbs energy change for the process in Fig. 1.16.

3. The third estimate was first published in Ben-Naim [3]. The process is the same as the one in Fig. 1.16, i.e. we bring two water molecules from infinite separation to the final configuration at which they *can* form a HB bridge by a water molecule. We emphasize again that at the final configuration the pair of water molecules are *solvated* by water molecules, i.e. the solvation Gibbs energy at this configuration is: $R = 4.5\,\text{Å}$, Ω_{HBB}

$$\Delta G^*(R = 4.5\,\text{Å}, \Omega_{HBB}) = -kT \ln \exp\left[-\beta B_{12}(R = 4.5\,\text{Å}, \Omega_{HBB})\right]_0 \tag{5.13}$$

The symbol Ω_{HBB} means that the two water molecules are in such a configuration that they *can* form a HB bridge (HBB) with a water molecule in the solvent.

For two molecules there are sixteen such configurations included in Ω_{HBB}. In Fig. 5.2 we show only one configuration, such that each water molecule is oriented so one of the OH direction is at angle α defined by:

$$\alpha = \frac{180 - \vartheta_T}{2} \approx \frac{180 - 109.5}{2} \approx 35^\circ \tag{5.14}$$

$\vartheta_T \approx 109.5$ is the tetrahedral angle, i.e., the angle between two pairs of O–O bonds in ordinary ice structure, Appendix A.

The total Gibbs energy change is written as (see Eq. 5.5):

$$\Delta G(R = 4.5\,\text{Å}, \Omega_{HBB}) = \Delta U + \delta G_{LJ} + \delta G_{HB|LJ} \tag{5.15}$$

at the distance of $R = 4.5\,\text{Å}$, we may neglect the first two terms on the right-hand-side of Eq. (5.15). Thus, we assume that the main contribution to the Gibbs energy change in this process is due to the solvent-induced effect of the "turning-on" the hydrogen bond capability of the two molecules given that the LJ interactions of 1 and 2 have already been "turned on."

To estimate the quantity $\delta G_{HB|LJ}$, we first write it as:

$$\delta G_{HB|LJ} = -kT \ln \frac{\exp\left[-\beta B_{12}^{HB}\right]_{LJ}}{\exp\left[-\beta B_{1}^{HB}\right]_{LJ}^{2}} \tag{5.16}$$

In the numerator on the right-hand side of Eq. (5.16) we have the *conditional* solvation Gibbs energy of the HB part of the interaction of the pair of water molecules at the final configuration given that the LJ interactions of 1 and 2 have already been "turned on." Similarly, the denominator in the conditional solvation Gibbs energy at the initial configuration.

Since in the process as shown in Fig. 1.16, the solvation of only two arms changes, we can take in the denominator of Eq. (5.16) the conditional solvation Gibbs energy of one arm of a water molecule. Thus, in writing Eq. (5.16) we implicitly assumed that the change in Gibbs energy is due to the change in the conditional solvation of two arms; from the initial configuration at which the two arms are independently solvated, to the final configuration when the solvation of the two arms are dependent. In this view, the denominator is Eq. (5.16) is related to the conditional solvation Gibbs energy of one arm of a water molecule. This quantity has been estimated from the theoretical arguments and experimental data, see Appendix B.

$$-kT \ln \exp\left[-\beta B_{1}^{HB}\right]_{LJ} = -2.25\ \text{kcal/mol} \tag{5.17}$$

Thus, what we need to estimate is the numerator in Eq. (5.16).

We first write the numerator on the right-hand-side of (5.16) explicitly:

$$\exp\left[-\beta B_{12}^{HB}\right]_{LJ} = \int Pr(\mathbf{X}^N|LJ) \exp\left[-\beta B_{12}^{HB}\right] d\mathbf{X}^N \tag{5.18}$$

Note that by B_{12}^{HB} we mean the HB *binding* energy of the two arms (molecules 1 and 2) at the final configuration in Fig. 5.2. The average in Eq. (5.18) is over all configurations of the N solvent molecules given that the LJ part of the interaction of the two molecules 1 and 2 with the N solvent molecules have been "turned on." (In application of Eq. (5.16) to proteins the two "arms" would be any of the arms—donor or acceptor—of a hydrophilic group such as OH, C=O or NH. The condition "LJ" would be replaced by the entire backbone of the protein at some specific confirmation).

In the integration over all configurations over $\mathbf{X}^N$, there are some configurations of all the water molecules where at most, one water molecule is HBed to either 1 or 2, or to both molecules 1 and 2. (This is the configuration we assumed in the previous estimate).

Therefore, we can split the total range of integration in Eq. (5.18) as follows:

$$\int Pr(X^N|LJ)\exp[-\beta B_{12}^{HB}]dX^N = \int_{\substack{one\,w\\ HBed\,to\,1}} \cdots dX^N + \int_{\substack{one\,w\\ HBed\,to\,2}} \cdots dX^N + \int_{\substack{one\,w\\ HBed\,to\,1,2}} \cdots dX^N + \int_{\substack{no\,w\,is\,HBed\\ to\,either\,1\,or\,2}} \cdots dX^N \quad (5.19)$$

The first integral is over all the configurations of the N water molecules, such that one water (w) molecule is HBed to the water molecule numbered 1. The second is the same but the w is HBed to 2. The third is over all configurations such that a water molecule is HBed to both 1 and 2, i.e. forms a HB-bridge, as in Fig. 5.2. The last integral is over all configurations for which no water (w) forms HB with either 1 or 2. For all of the configurations in the last term in Eq. (5.19) the HB-binding energy B_{12}^{HB} is zero. Therefore, the last term on the right-hand-side of (5.19) is simply the probability of the event: "No water molecule is HBed to either 1 or 2. We write this explicitly as:

$$\int_{\substack{now\,is\\ HBed\,to\\ either\,1\,or\,2}} Pr(X^N|LJ)dX^N = Probabiliy\ that\ no\ water\ molecules\ is\ HBed\ to\ either\ 1\ or\ 2 \quad (5.20)$$

This probability is equal to one minus the probability of the complimentary event, i.e. the event that at least one water molecule forms a HB with either 1 or 2, or both.

$$\int_{\substack{now\,is\\ HBed\,to\\ either\,1\,or\,2}} Pr(X^N|LJ)dX^N = 1 - \int_{\substack{one\,w\\ HBed\,to\\ 1}} \cdots dX^N - \int_{\substack{one\,w\\ HBed\,to\\ 2}} \cdots dX^N - \int_{\substack{one\,w\\ HBed\,to\\ 1\,and\,2}} \cdots dX^N \quad (5.21)$$

Note that in the original publication Ben-Naim [3] the range of integrations in Eq. (5.19) was over "*at least one water molecule forms* HB with…." However, since only *one* water molecule can form a HB with either 1 or with 2, or HBB with 1 and 2, the range of the integrations, on the first three integrals on the right-hand-side of (5.19) was changed into "one water molecules HBed to…".

Combining Eq. (5.19) with (5.21) we obtain:

$$\int Pr(X^{W}|LJ)\exp[-\beta B_{12}^{HB}]dX^{N} =$$

$$1 + \int\limits_{\substack{one\,w\\ HBed\,to\\ 1}} Pr(X^{N}|LJ)\{\exp[-\beta B_{12}^{HB}] - 1\}dX^{N}$$

$$+ \int\limits_{\substack{onew\\ HBedto\\ 2}} \cdots dX^{N} + \int\limits_{\substack{onew\\ HBedto\\ 1and2}} \cdots dX^{N} \tag{5.22}$$

The ellipsis "…" means the same integrand as the one in the preceding integral, Now, each of the integrals on the right-hand-side of (5.22) may be further simplified as follows:

$$\int\limits_{\substack{one\,w\\ HBed\,to\,1}} \cdots dX^{N} = \sum_{i=3}^{N+2}\int Pr(X_i|LJ)\{\exp[-\beta B_{12}^{HB}] - 1\}dX_i$$

$$= N\int Pr(X_3|LJ)\{\exp[-\beta B_{12}^{HB}] - 1\}dX_3$$

$$= \int \rho(X_W|LJ)\{\exp[-\beta B_{12}^{HB}] - 1\}dX_W \tag{5.23}$$

In the first step on the right-hand-side of (5.23) we have integrated over all the configurations of the N water molecules. Since the two water molecules at the fixed configurations were denoted by 1 and 2, we number all the other "solvent" molecule by $i = 3$ to $i = N + 2$. This integration is possible since there is only *one* water molecule that can form a HB. Thus, the entire integral on the left-hand-side is split into N integrals, in each of which one specific water molecule (denoted i) forms the BH bridge.

The resulting integrand is the probability density of finding a water molecule (i) at a configuration X_i, given two water molecules, 1 and 2 at fixed configurations as in Fig. 5.2, and that only the LJ part of the interaction has been "turned on."

Next, we recognize that each of the N terms in the sum have equal value. This follows from the equivalency of all the N water molecules. Therefore, we choose one specific water molecule, say $i = 3$ and replace the sum over all N equal terms by N times one integral.

In the final step we replace the *specific* probability $Pr(X_3|LJ)$ by the *generic* probability density $\rho(X_w|LJ)$ which can also be interpreted as the *local density* of water molecules at X_W given that there are two water molecules 1 and 2 at fixed configurations and interact with the solvent through LJ interaction.

Similar simplification applies to all the three integrals on the right-hand-side of Eq. (5.22). We now recognize that when we integrate over all possible configurations of the water (w) molecule, there is only a small range of configurations for which the integrand is non-zero (note that the binding energy B_{12}^{HB} is due only to HBing, therefore, for any configuration X_W for which the water molecule does not form a HB with either 1 or 2, B_{12}^{HB} will be zero, hence, also the integral in (5.23) will be zero.

Denoting by ΔX_{HB} the region from which a water molecule can form a HB with either 1 or 2, and by ΔX_{HBB} the region from which a water molecule can form a HB-bridge with 1 and 2, and assuming that in these regions the value of the integrand is constant we can rewrite the approximate form of Eq. (5.22), noting (5.23) as:

$$\begin{aligned}\int Pr(\mathbf{X}^W|LJ)\exp[-\beta B_{12}^{HB}]d\mathbf{X}^N &= 1\\ &+ 2\{\exp[-\beta\varepsilon_{HB}] - 1\}2\rho_W g_1\frac{\Delta X_{HB} - \Delta X_{HBB}}{8\pi^2}\\ &+ \{\exp[-\beta 2\varepsilon_{HB}] - 1\}2\rho_W g_2\frac{\Delta X_{HBB}}{8\pi^2}\end{aligned} \qquad (5.24)$$

On the right-hand-side of Eq. (5.24), the second term is due to the two terms in Eq. (5.22) corresponding to the formation of one HB (to either 1 or 2). The conditional density is written as the bulk density of water times the pair-correlation function at that configuration which we denoted by g_1. The third term corresponds to the case where a water molecule can form a HBB with 1 and 2, and we denoted the corresponding pair correlation function by g_2.

The range of configurations ΔX_{HB} was estimated in Appendix B in connection with the calculation of the solvation Gibbs energy of water in water. g_1 and g_2 were arbitrarily chosen to be of the order of 2 and ρ_W was taken as the bulk density of liquid water at 25 °C, and ε_{HB} as the HB energy $\varepsilon_{HB} = -6.5$ kcal/mol. Note than on the second term on the right-hand-side of Eq. (5.24) we have one HB energy (ε_{HB}), whereas in the third term we have two HB energies ($2\varepsilon_{HB}$).

With all of these admittedly crude approximations we estimated the Gibbs energy change for the process in Fig. 1.16 to be about:

$$\Delta G(R = 4.5\,\text{Å}, \Omega_{HBB}) \approx -3\,\text{kcal/mol} \qquad (5.25)$$

It should be emphasized that in the 1980s no one ever mentioned the possibility of such $H\phi I$ effect (recall that at that time the *direct* HB was deemed to be insignificant in the stability of proteins). The value estimated in Ben-Naim [3], was admittedly very approximate, but it was similar to other values I estimated based on experimental data, which I will described in the next section. What was the most exciting discovery (at least for me, at that time), was the fact that there could be other SII which were stronger than the $H\phi O$ effect. This was the beginning of a mini-revolution in the field of protein folding as well as in other biochemical problems discussed in Chap. 6.

5.2 Evidence for $H\phi I$ Effect Based on Experimental Data

One experimental piece of data we mentioned in Sect. 5.1 was the location of the second peak in the radial distribution function at a distance of $R = 4.5$ Å. This distance is the same as the distance between the second nearest-neighbor oxygen atoms in the structure of ordinary ice, see Appendix A. This fact indicates that in liquid water, two water molecules are more likely to be found at a distance of $R = 4.5$ Å than any of the distances in the range of say, $3 \lesssim R \lesssim \infty$. The only other, more probable distance is around $R \approx 2.8$ Å. The rational for the relatively large probability of finding a pair of water molecules at a distance of 4.5 Å was explained in Sect. 5.1. The argument is based on the relatively high probability of finding a triplet of water molecules connected by two HBs, as shown in Fig. 5.2.

Soon after I had made the rough estimate of the strength of the $H\phi I$ interaction at a distance of $R = 4.5$ Å, see Sect. 5.1, I search the literature to look for any supporting evidence for my estimate.

The only relevant data I found was an article by Haberfield et al. published in 1984 [4]. In this article, Haberfield et al. reported on the distribution coefficients of some compounds between water and toluene, see Fig. 5.5 and Table 5.1.

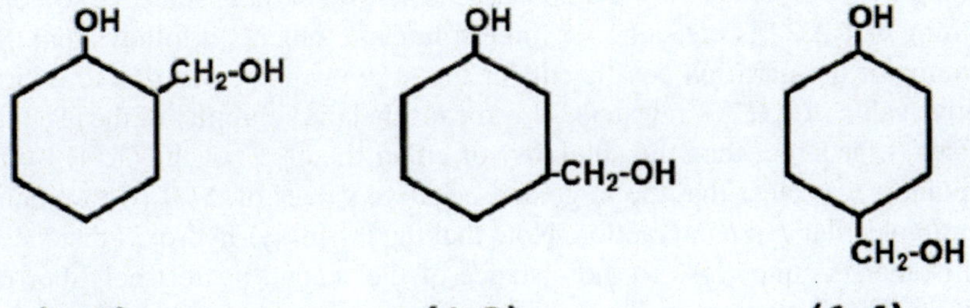

(1.2) (1.3) (1.4)

Fig. 5.5 The three isomers of the same compound, for which experimental data on the distribution coefficients between water and toluene were available

Table 5.1 The distribution coefficient $P_d(toluene - water)$ and Gibbs energy change for the transfer (toluene → water) for the three derivatives of phenol at 25 °C

Compound	Pd(toluene – water)	ΔG(toluene – water)
(1.2)	0.13	– 1.20 kcal/mol
(1.3)	0.0089	– 2.80 kcal/mol
(1.4)	1.79	0.35 kcal/mol

What was the most striking to me in reading these results was that the partition coefficient of the (1, 3) isomer was more than two ordered magnitudes smaller than the values for (1, 2) and the (1, 4) isomers. Even before analyzing these results I had the intuitive feeling that this striking result must be related to what I called the intramolecular $H\phi I$ interaction.

The partition coefficient is related to the Gibbs energy of transfer as follows:

$$P_d(toluene - water) = \frac{\rho_s(toluene)}{\rho_s(water)}$$
$$= \exp[-\beta\Delta G_s(water \to toluene)$$
$$= \exp\left[-\beta\Delta G_s^*(in\,toluene) + \beta\Delta G_s^*(in\,water)\right) \tag{5.26}$$

Here, ρ_s is the concentration of the solute s, and ΔG_s^* is the solvation Gibbs energy of s in the appropriate solvent. The question is why the (1.3) isomer has a partition coefficient much smaller than either the (1, 2) or the (1, 4) isomer.

It should be noted in the old literature on thermodynamics of solvation, it was common to assume that the solvation Gibbs energy is "additive," meaning that the value of ΔG_s^* of the solute s is the sum of the solvation Gibbs energies of all the *groups* comprising the solute s. If that assumption is true, then we should expect that the solvation Gibbs energies of the three isomers in Fig. 5.5 would be approximately the same. The fact that the partition coefficient is so different for the three isomers was initially attributed to the solvation Gibbs energies of these solutes in water. Thus, assuming that ΔG_s^* is the same for three solutes in toluene, it follows that the very low value of the partition coefficient for the (1, 3) isomer, was due to a large and negative value of $\Delta G_{1,3}^*$ (in water). This means that the solubility of the (1, 3) isomer in water is far larger than the solubility of either the (1, 2) or the (1, 4) isomer. At this point it was clear that the large and negative values of $\Delta G_{1,3}^*$(in water) is due to intramolecular $H\phi I$ interaction. Note that the Hydroxyl-hydroxyl distance in the (1.3) isomer is quite close to the distance of the second nearest-neighbor oxygen atoms in the structure of ordinary ice, see Appendix A, which is $R = 4.5$ Å, see Fig. 5.6.

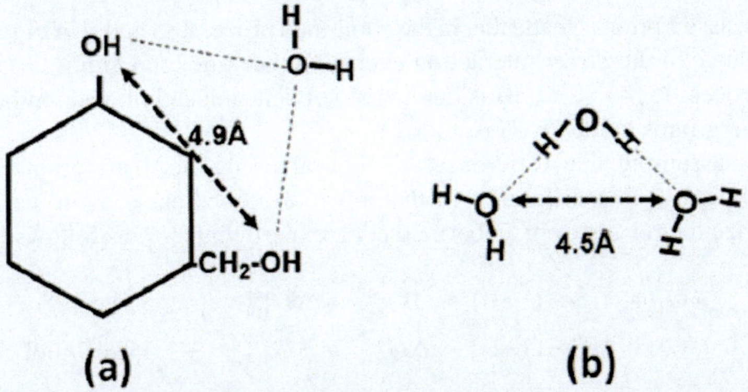

Fig. 5.6 **a** The distance between the two hydroxyl groups in the (1.3) isomer in Fig. 5.5, is slightly larger than the distance of the second nearest neighbors of oxygen atoms in ordinary ice **b**

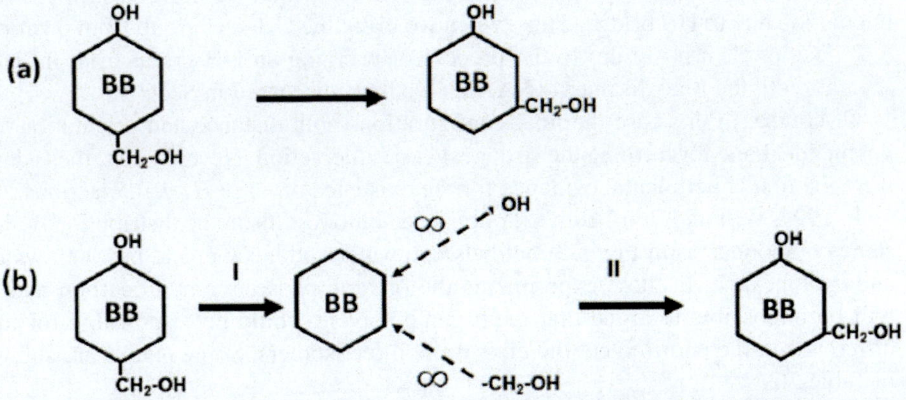

Fig. 5.7 **a** The process of transferring a group from position "4" to position "3". **b** The same as in **a**, but the process is carried out in two steps I and II

To understand the large value of the solvation Gibbs energy of the (1.3) isomers, consider the following hypothetical process shown in Fig. 5.7a. We start with the (1, 4) isomer at some fixed configuration in water. The two hydrophilic groups may be considered to be independently solvated (they are far apart from each other). We next "cut" the two $H\phi O$ groups, move them to "infinite" distance, and then combine them to form the new isomer (1, 3). We can write the total Gibbs energy change for this process is:

$$\Delta G[(1,4) \to (1,3)] = \Delta U[(1,4) \to (1,3)] + \delta G[(1,4) \to (1,3)] = \Delta G^*_{1,3} - \Delta G^*_{1,4} \tag{5.27}$$

Thus, assumption is made that in the combined processes I and II in Fig. 5.7b, there is no change in the *direct* interaction energy. Hence, the total Gibbs energy change in the process $(1, 4) \rightarrow (1, 3)$ is due to the solvent-induced interaction between the two $H\phi I$ groups in the $(1, 3)$ isomer.

Thus, assuming that differences in the values of the distribution coefficients is mainly due to the difference in the solvation Gibbs energies in water, we can summarize the results from Haberfield et al. experiments [4], as follows:

$$\delta G[(1,4) \rightarrow (1,3)] \approx \Delta G_{1,3}^{*W} - \Delta G_{1,4}^{*W} \approx -3.15\,\text{kcal/mol}$$
$$\delta G[(1,4) \rightarrow (1,2)] \approx \Delta G_{1,2}^{*W} - \Delta G_{1,4}^{*W} \approx -1.55\,\text{kcal/mol} \qquad (5.28)$$

The explanation for the large and negative Gibbs energy difference, is that in the $(1, 3)$ isomer the $O - O$ distance is about 4.9 Å (compare with a distance of about 2.2 Å and at 6.3 Å for the $(1, 2)$ and the $(1, 4)$ isomers, respectively).

The distance of 4.9 Å is very close to the 4.5 Å at which we expect a strong $H\phi I$ interaction due to HB bridge. Thus, when we bring the CH_2OH group from position 4 to position 3, it is similar to the process of bringing an OH group from infinite separation to the final distance 4.5 Å, that we have discussed in Sect. 5.1.

Of course, in this case the final configuration (both distance and orientation) is not the ideal one for forming the strongest $H\phi I$ interaction. Nevertheless, these data were the first experimental evidence for the existence of large $H\phi I$ interactions.

In 1994, Wilf and Ben-Naim [5] published some results on the distribution of the same compounds as in Fig. 5.5, but between water and hexane, and between water and cyclohexane. In these experiments the organic phase is a hydrocarbon and it will be impossible to argue that the organic solvent would be responsible for the difference in the partition coefficients of the three isomers. Some results are shown in Table 5.2.

Again, we see that transferring the group CH_2OH from either position 4, or position 2 to position 3 involves a negative solvent-induced free energy change. It is clear that transferring from position 2 to position 3 involves the largest SII.

More recently similar experiments on the distribution of the three isomers of di-hydroxybenzene were done. Lazaric et al. (to be published, 2023).

Again, it is found that transferring a hydroxyl group from position 4 to position 3 involves a large solvent-induced interaction of about − 3 kcal/mol.

Table 5.2 Values of $\delta G(4 \rightarrow 3)$, $\delta G(4 \rightarrow 2)$, $\delta G(2 \rightarrow 3)$ from the experiments on the distribution of the three solutes in Fig. 5.5, in hexane-water and cyclohexane-water, in kcal/mol and at 25 °C (From Wilf and Ben-Naim [5])

	Hexane-water	Cyclohexane-water
$\delta G(4 \rightarrow 3)$	− 1.28	− 1.61
$\delta G(4 \rightarrow 2)$	+ 1.00	+0.92
$\delta G(2 \rightarrow 3)$	− 2.28	− 2.53

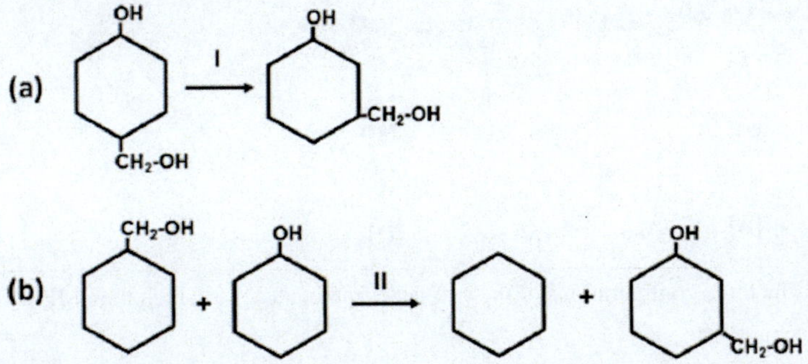

Fig. 5.8 **a** The same process as in Fig. 5.7a. **b** A modified process having similar SII as in process **(a)**

Regarding the assumption that the two $H\phi I$ groups in the (1, 4) isomer are independently solvated, we compared the following two reactions:

Process I is the same as in Fig. 5.7a, and is shown in Fig. 5.8a. In the new process, Fig. 5.8b, we take two molecules at infinite separation in the solvent and transfer the group CH_2OH from one molecule to the second molecule at position 3. We get the (1, 3) isomer and the benzene molecule. The indirect SII in this case is calculated from:

$$\delta G(II) = \Delta G^*_{Benzene} + \Delta G^*_{1,3} - \Delta G^*_{B-OH} - \Delta G^*_{B-CH_2OH} \tag{5.29}$$

The values obtained for the "reactions" I and II were very similar. The results support the assumption that the distance between the two $H\phi I$ groups in the (1, 4) isomer is effectively "infinity," i.e. the two $H\phi I$ groups are independently solvated.

Very recently the solvation Gibbs energies of the three isomers of Benzene di-carboxylic acid were calculated. These calculations are consistent with the experimental results on the solubility of the isomers of Benzene di-carboxylic acid. Unlike the isomers in Fig. 5.5, in this case, the solubility of the 1, 2 isomers is the largest (indicated strongest solvation Gibbs energy) compared with the 1, 3 and the 1, 4 isomers.

Finally, another experimental fact which is consistent with the strong SII is the very large difference in the solubility of the two isomers: Fumaric and Maleic acids. We shall discuss that in the next chapter.

5.3 Simulated Results on $H\phi I$ Interactions and Forces

Immediately after the first theoretical argument, published (1989) [3], and followed by the confirmation, based on experimental results, I felt it is necessary to get also some simulated results on the $H\phi I$ interactions. I asked my friend and colleague

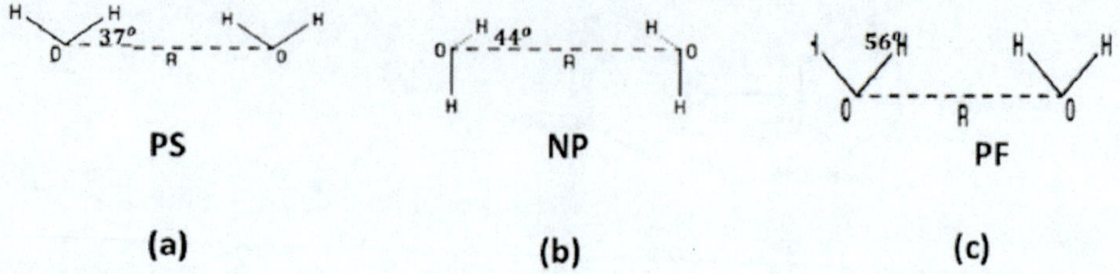

Fig. 5.9 The three configurations of the two water molecules, for which the PMF was calculated, see Table 5.3

Mihaly Mezei to do simulation calculations of the potential of mean force (PMF) between two water molecules at fixed configurations.

In 1990, Mezei and Ben-Naim [6], published the first Monte-Carlo computer simulations on a sample of 216 water molecules in which two water molecules are held at fixed configurations. These configurations are shown in Fig. 5.9.

The first configuration (PS), is shown in Fig. 5.9a. The two water molecules (i.e. the centers of the oxygen and hydrogen atoms) are in the XZ plane, and the OH bonds form an angle of 37° with respect to the x-axis. The second configuration (NP), is shown in Fig. 5.9b. Here, the OH bonds form an angle of 44° with the x-axis. The third configuration (PF), is shown in Fig. 5.9c.

For these three configurations, the solvent contribution to the PMF was calculated. Table 5.3 shows some of the results from these calculations. In general, these results were consistent with both the theoretical and the experimental results, namely that $H\phi I$ interactions are far stronger than the corresponding $H\phi O$ interactions.

Note that in all of the configurations, shown in Fig. 5.9, a HB-bridge can be formed between the two water molecules at a distance of 4.5 Å. Note that in the third configuration (c), two water-bridges may be formed symmetrically above and below the x-axis.

The result of this work clearly supported the theoretical findings from Ben-Naim [3], and the estimates based on experimental results discussed in Sect. 5.2. At this stage it was clear that the "myth" that $H\phi O$ interactions are the most important factor in the stability of the proteins had collapsed. Instead, there was overwhelming evidence in favor of the importance of the $H\phi I$ interaction. This evidence did not address the question regarding the speed of the folding process. To answer this

Table 5.3 The solvent contributions to the PMF for the three configurations as shown in Fig. 5.9

Configuration	Water model	Solvent contribution to PMP in kcal/mol
(a)	TIP4P	– 3.2
(a)	SPC	– 3.9
(b)	TIP4P	– 1.7
(c)	TIP4P	– 6.0

question, one needs to know the *forces* exerted on the different groups along the protein (see Sect. 6.2).

In 1994, immediately after the recognition of the importance of the *forces* to solving the speed of the folding problem, Durell et al. [7], published the first simulated results on the solvent-induced forces on both $H\phi O$ and $H\phi I$ groups. The calculations were based on the formula for the solvent-induced forces published earlier in Ben-Naim [2]. The molecular dynamic calculations were performed using the TIPS3P model for water. The configuration used for these simulations is essentially the one shown in Fig. 5.2, see also Eq. (5.14).

Figure 5.10 shows the solvent-induced forces computed from equation:

$$\boldsymbol{F}_1^{(SI)} = -\int \nabla_1 U(\boldsymbol{X}_1, \boldsymbol{X}_W)\rho(\boldsymbol{X}_W|\boldsymbol{X}_1, \boldsymbol{X}_2)d\boldsymbol{X}_W \tag{5.30}$$

Note that the strongest force is found at the distance of about 4.9 Å.

By "switching-off" the charge-charge interactions in the TIPS3P model we get a Lennard–Jones particle. Thus, repeating the same calculations for these LJ particles we get the corresponding solvent-induced force on particle 1 given the two LJ particles at the same distance.

It was found that the maximum attractive forces on the water molecule is more than twice larger than the force on the LJ particle.

By integrating the solvent-induced force, Durell et al. [7], also calculated the solvent-induced PMF. These calculations confirm the earlier results by Mezei and Ben-Naim [6], that the solvent-induced part of the PMF between $H\phi I$ molecules is much larger than $H\phi O$ molecules.

A more systematic study of the forces on $H\phi O$ and $H\phi I$ solutes was undertaken by Durell and Ben-Naim [8].

We first show some results on the total force on a $H\phi O$ solute, given another $H\phi O$ solute at a distance R (in Å), Fig. 5.11.

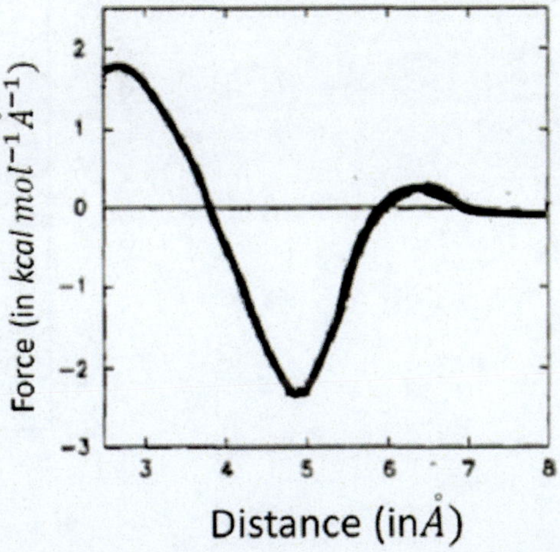

Fig. 5.10 The first simulated results on solvent-induced hydrophilic forces between two water molecules at the configuration as in Fig. 5.2. See Eq. (5.30)

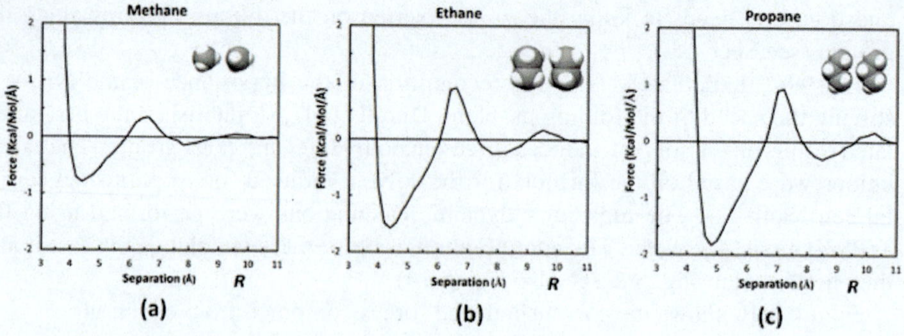

Fig. 5.11 Simulated results on the force on one hydrophobic molecule (**a** methane, **b** ethane, and **c** propane), given another hydrophobic molecule at a distance R

The three solutes are methane, ethane, and propane. The strongest attractive force between the two methane molecules is, at a distance of about $R \approx 4.5$ Å, and its value is about $-0.8\,\text{kcalmol}^{-1}\,\text{Å}^{-1}$. For ethane it is about $-1.6\,\text{kcalmol}^{-1}\,\text{Å}^{-1}$, and for propane and isobutane, the values are about the same, $-1.9\,\text{kcalmol}^{-1}\,\text{Å}^{-1}$. (Note that in the original article by Durell and Ben-Naim [8], in Fig. 5.4 there is an error: "pentane" should be replaced by "propane," in both the figure and in the caption).

We now show a few results on hydrophilic forces.

First, we show the hydrophilic forces between two water molecules that can form direct hydrogen bond. The configuration is shown in Fig. 5.1, and the total force as well as direct and indirect components are shown in Fig. 5.12.

Fig. 5.12 Simulated results on the force on one water molecule given another water molecule at a configuration as in Fig. 5.1, as a function of the distance R

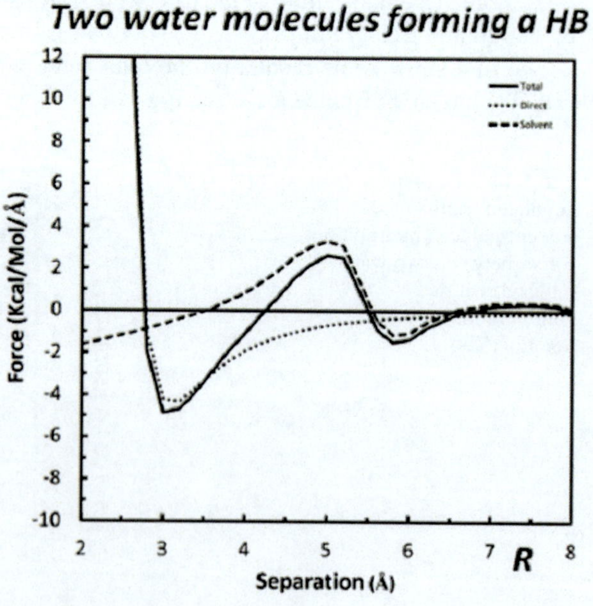

Note that the total force has a minimum (i.e. maximal attractive force) at the distance of $R = 3$ Å. This is slightly larger than the distance of 2.8 Å in which the radial distribution function has a maximum.

In Fig. 5.13 we show the hydrophilic force between two water molecules at the configuration shown in Fig. 5.2, i.e., such that a HB bridge can be formed. Figure 5.13 shows the total force, the direct and the indirect contributions to the force, as a function of the distance between the two water molecules being at fixed orientation as in Fig. 5.2. The maximal total force is at about $R = 5$ Å. It should be noted that this curve is the *total* force both direct and indirect forces. The indirect part is, in fact, stronger at $R = 5$ Å than the total force. The direct part is repulsive in this range of distances due to the electrostatic repulsion between the partial charges on the oxygen and hydrogen atoms.

Figure 5.14 (right) shows the configuration of three water molecules. Molecule 3 approaches 1 and 2 at such a configuration in which it can form simultaneous two direct HBs with both 1 and 2. The force curves are shown in Fig. 5.14 (left). We see here that the total force is much stronger with a minimum of about -8.33 kcalmol^{-1} Å^{-1} at a distance of about 2.2 Å (this distance is the one in which water number 3 can form two direct HBs with 1 and 2. Note that the largest contribution to the *total* force comes from the *direct* HBs.

Next, we show the case when one water molecule approaching a pair of water molecules 1, 2 at a fixed configuration as shown in Fig. 5.15 (right). The force curve is shown in Fig. 5.15 (left). In this case the direct and the indirect forces contribute negatively (i.e. attractive) to the total force. The maximal total attractive force is

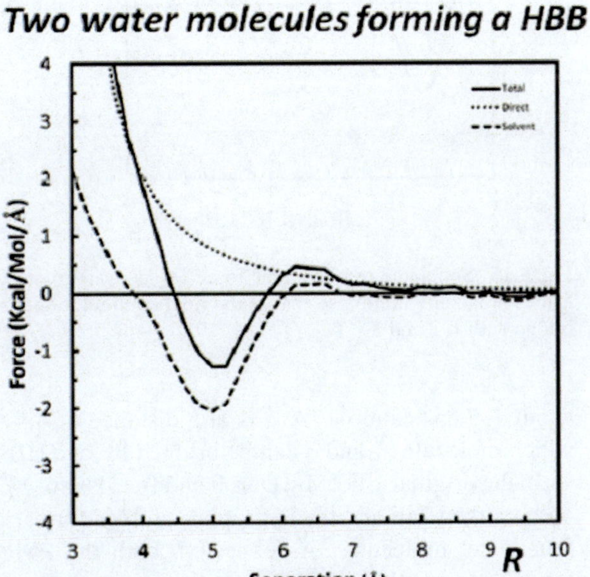

Fig. 5.13 Simulated results on the force on one water molecule given another water molecule at a configuration as in Fig. 5.2, as a function of the distance R

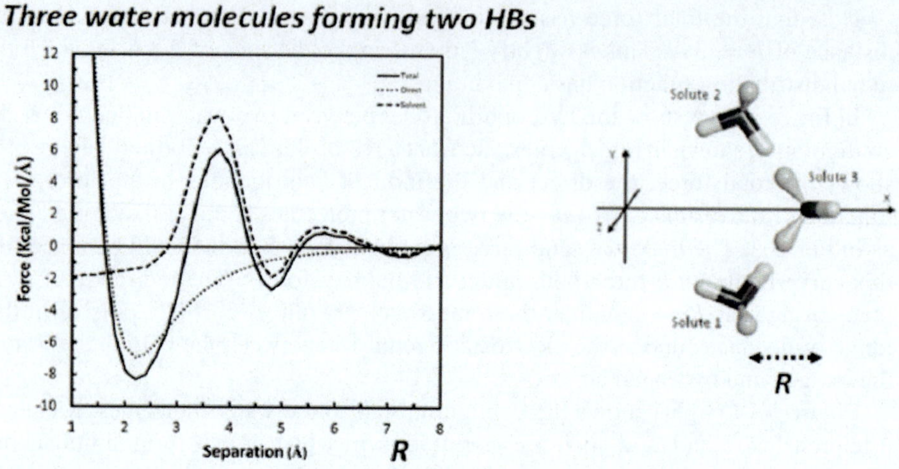

Fig. 5.14 Simulated results on the force on one water molecule, numbered (3), given two other molecules at a configuration such that it can form two HB with 1 and 2

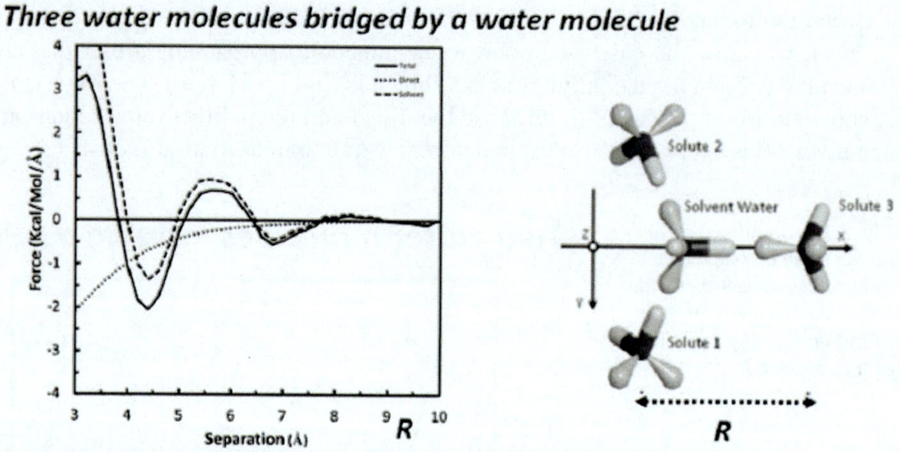

Fig. 5.15 Simulated results on the force on one water molecule (3), given two other molecules (1 and 2) at a configuration so that at the final configuration a solvent molecule can form a 3-water bridge with 1, 2 and 3

about $-2.05\ \text{kcalmol}^{-1}\ \text{Å}^{-1}$ is at a distance of 4.4 Å (This is the same distance at which molecules 1 and 3 can be bridged by two HBs).

In the original article by Durell and Ben-Naim [8], much more data was presented on both the PMF and the hydrophilic forces for various configurations of three and four water molecules. As expected, both the PMF and the force become larger for greater number of water molecules. And as we have seen for two particles the hydrophilic forces are always stronger than the corresponding force between hydrophobic particles.

5.4 Temperature Dependence of $H\phi I$ Interactions and Forces

In Sect. 4.4 we discussed the temperature dependence of the $H\phi O$ interactions. Recently, Durell and Ben-Naim [9], published a extensive study of the temperature dependence of Hydrophobic-Hydrophilic interactions and forces. We found that the PMF between two $H\phi O$ molecules becomes *stronger* (i.e. more negative) when the temperature increases. This is true for temperatures around room temperature.

Figure 5.16a shows the minimum value of the PMF for two methane and three methane molecules as a function of temperature between 280 and 360° K. The dependence on temperature is almost linear, and the negative slopes of these curves was interpreted in terms of structural changes induced in the solvent, see Sect. 4.4. Figure 5.16b shows the maximum attractive forces on one methane molecule given that one and two other methane molecules are at a fixed configuration (in which the maximum force is attained). We note that the maximum force reported here consists of both the direct and the indirect (i.e. solvent-induced) parts. Table 5.4 also shows the values of ΔG, ΔH and $T\Delta S$ for the process of bringing two and three methane molecules from infinite separation to the final configuration in which PMF is minimum, denoted "two-methane" and "three- methane," respectively, see inserts in Fig. 5.11. The entropy and the enthalpy changes for these processes are calculated from the standard equations:

$$\Delta S = -\frac{\partial \Delta G}{\partial T}, \text{ and } \Delta H = \Delta G + T\Delta S \tag{5.31}$$

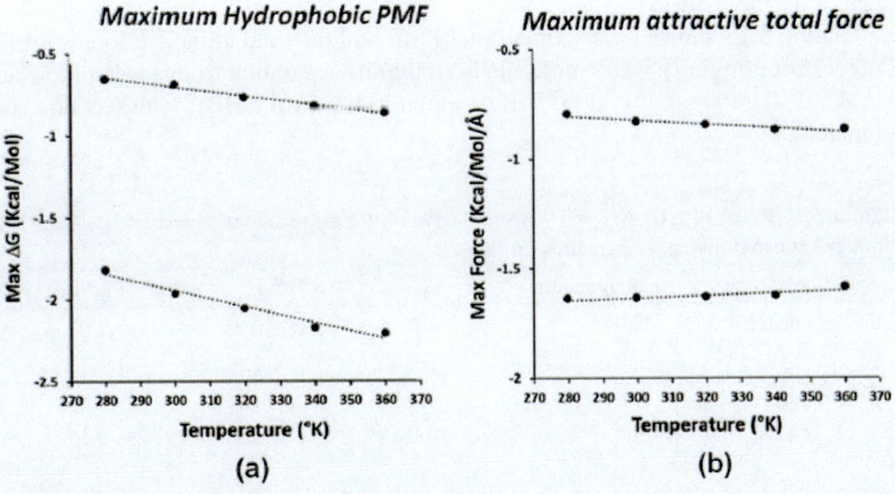

Fig. 5.16 Temperature dependence of the **a** maximum PMF, and **b** the total maximum attractive-force between 2 methane molecules (upper curve) and 3 ethane molecules (lower curve)

Table 5.4 Values of ΔG, ΔH and $T\Delta S$ (in kcal/mol) for the reactions as specified at two temperatures (in $^{\circ}K$), see inserts in Fig. 5.11

Reaction	Temperature	ΔG	ΔH	$T\Delta S$
Two-methane	280	−0.6	+0.2	+0.8
	360	−0.8	+0.2	+1.0
Three-methane	280	−2.5	+0.5	+3.0
	360	−3.0	−1.2	+1.8

Note that the temperature dependence of ΔG, ΔH and $T\Delta S$ is relatively small for the "two-methane" compared to the case for "three-methane."

Next, we consider the reaction "direct one HB" shown in Fig. 5.1. In this reaction we bring two water molecules to the final distance and configuration in which they form *direct* HB. The second process is shown in Fig. 5.2 in which the two water molecules are brought to the final configuration in which they can form HB bridge by a "solvent" water molecule (Table 5.5).

One of the most important aspects of the temperature dependence of these two $H\phi I$ interactions is that the first has a negative slope (i.e. the strength of the $H\phi I$ becomes stronger) similar to the case of $H\phi O$ interaction (see Table 5.4). The second has a positive slope (i.e. the $H\phi I$ interaction becomes weaker as the temperature increases). This opposite temperature dependence was assumed earlier to explain the curious phenomenon of "cold denaturation" of proteins. We shall discuss this topic in the next chapter.

Figure 5.17 shows the maximal total PMF and the total attractive forces for the process of bringing a water molecule from infinite separation from a water molecule to the final configuration at which it can a direct HBs with second water molecules (Fig. 5.1; Table 5.6).

Figure 5.18 shows the maximal total PMF and the total attractive forces for the process of bringing a water molecule from infinite separation from a water molecule to the final configuration (Fig. 5.2) in which it can form a HBB with second water molecules.

Table 5.5 Values of ΔG, ΔH and $T\Delta S$ (in kcal/mol) for the reactions "direct one HB" and "HB bridge" at two different temperatures (in $^{\circ}K$)

Reaction	Temperature	ΔG	ΔH	$T\Delta S$
Direct one HB	280	−2.75	−0.75	+2.0
	360	−3.40	−2.6	0.8
HB-bridge	280	−0.75	−1.25	−0.5
	360	−0.55	−1.5	−0.15

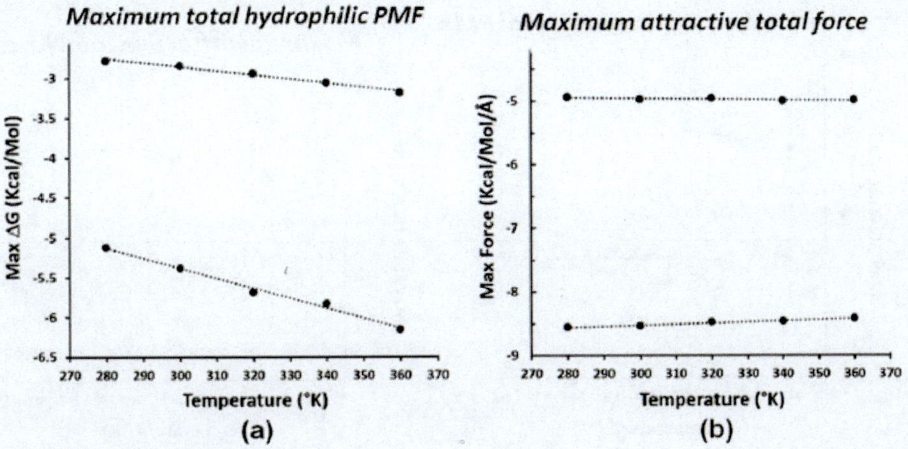

Fig. 5.17 Temperature dependence of the **a** maximum total PMF and **b** The maximum total attractive force for the configuration at which they form a direct HB, as in Fig. 5.1. Lower curves for two water molecules and upper curves for three water molecules

Table 5.6 Values of ΔG, ΔH and $T\Delta S$ (in kcal/mol) for the reactions "direct two HB" and "three HB bridge" at two different temperatures (in $^\circ K$), based on data from Fig. 5.8 of Durell and Ben-Naim [9]

Process	Temperature	ΔG	ΔH	$T\Delta S$
Direct two HBs	280	− 7.9	− 3.4	− 4.4
	360	− 9.3	− 2.4	+ 6.8
3 HB-bridge	280	− 2	− 2.7	− 0.6
	360	− 1.8	− 2.8	− 1

5.5 Concluding Remarks

In 1959, Kauzmann published a review article on "Some factors in the interpretation of protein denaturation" [10]. In this article Kauzmann introduced the idea that HϕO groups (or molecules) "*tend to adhere to each other in aqueous solutions*." This tendency was originally called "hydrophobic bond." Later, the term evolved into HϕO interaction, and HϕO solvation.

In the same article, Kauzmann concluded that the formation of HBs between two HϕI groups on proteins cannot contribute significantly to the stability of proteins.

For over fifty years, Kauzmann's ideas have had two major effects on biochemical researchers. First, people in the field of science considered the HϕO effect as a "universal" explanation to many biochemical processes. Second, no one searched for other possible solvent-induced effects. The HϕO effect stood alone as the powerful explanation of everything.

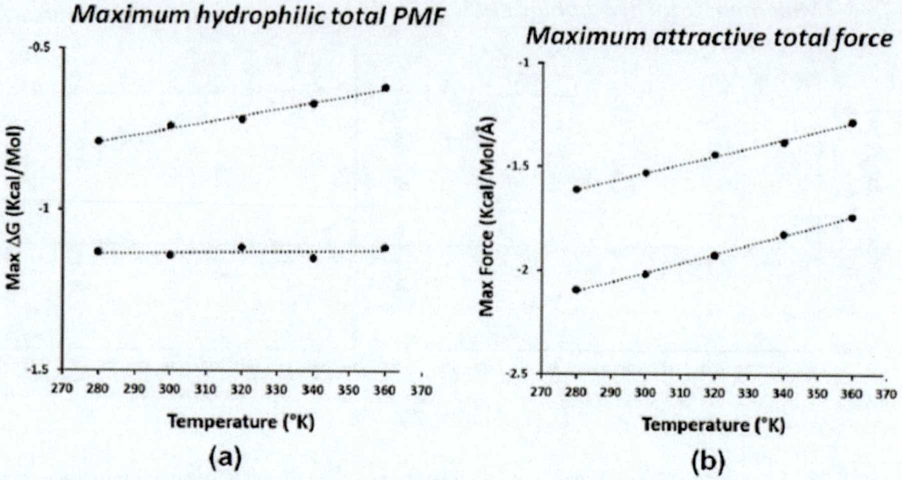

Fig. 5.18 Temperature dependence of the **a** maximum total PMF and **b** The maximum total attractive force for the configuration of HBB, as in Fig. 5.2. Lower curves for two water molecules and upper curves for three water molecules

A major turning point occurred in the early 1990s when a systematic study of *all possible* solvent-induced effects was undertaken. It was found that Kauzmann's reasoning for both the role of the HϕO effect, and for the insignificant contribution of the direct HBs were wrong. Instead, it was discovered that certain solvent-induced interactions and forces involving HϕI groups are far stronger than the corresponding HϕO effects. During the last 30 years theoretical arguments, experimental results and simulation calculations have supported the conclusion that HϕI interactions and forces are probably more important than the HϕO effects in many biochemical processes. Some of them will be discussed in the next chapter.

References

1. Ben-Naim A (2009) Molecular theory of water and aqueous solutions, part i: understanding water. World Scientific, Singapore
2. Ben-Naim A (2006) Molecular theory of solutions. Oxford University Press, Oxford, UK
3. Ben-Naim A (1989) Solvent-induced interactions: hydrophobic and hydrophilic phenomena. J Chem Phys 90:7412–7525
4. Haberfield P, Kivuls J, Haddad M, Rizzo T (1984) Enthalpies, free energies, and entropies of transfer of phenols from nonpolar solvents to water. J Phys Chem 88:1913–1916
5. Wilf J, Ben-Naim A (1994) A direct measure of hydrophilic interaction. J Phys Chem 98:8594–8595
6. Mezei M, Ben-Naim A (1990) Calculation of the solvent contribution to the potential of mean force between water molecules in fixed relative orientation in liquid water. J Chem Phys 92:1359–1361
7. Durell SR, Brooks BR, Ben-Naim A (1994) Solvent induced forces between two hydrophilic groups. J Phys Chem 98:2198–2202

8. Durell SR, Ben-Naim A (2017) Hydrophobic-hydrophilic forces in protein folding. Biopolymers 10:1–14
9. Durell, S.R. and Ben-Naim A (2021) Temperature dependence of hydrophobic-hydrophilic forces. J Phys Chem B 125:13137–13146
10. Kauzmann W (1959) Some factors in the interpretation of protein denaturation. Adv Protein Chem 14:1–63

Chapter 6
Implementation of the $H\phi I$ Interactions and Forces in Some Biochemical Processes

In Chap. 1 we surveyed the evolution of the ideas that were suggested to understand the so-called protein folding problem (PFP). Briefly, the question regarding the factors that are responsible for the stability of the 3D structure of proteins was initially answered soon after the structures of various proteins were discovered. It was found that large parts of the secondary structure of the protein—the α-helix and the β-sheet—are characterized by repeated intra-molecular HBs. It was natural (and quite correct) to conclude that intramolecular HBs are also the main factors that are responsible for maintaining the stability of the 3D structure of the proteins.

As we explained in Chap. 1, as a result of misinterpretation of some experimental results by John Schellman, people wrongly concluded that intramolecular HBs are insignificant in maintaining the stability of proteins. At that point, a kind of intellectual "vacuum" was created in the field of protein chemistry. Interestingly, Walter Kauzmann, the person who created this vacuum was quick to suggest an alternative factor to explain the stability of proteins. This is how the great myth of the importance of the $H\phi O$ effect was born and dominated the whole field for over half of a century.

It is not uncommon to find even in very recent textbooks on biochemistry statements claiming that $H\phi O$ effect is the "most important" driving force for protein fording. Some quotations were presented in Chap. 1, Sect. 1.4.

The main argument that was given by Kauzmann [1], and repeated by many authors [2] is that the free energy of transferring a non-polar solute from water to a non-aqueous solvent is large and negative. This fact together with the known fact that most of the hydrophobic residues of a protein are found in its interior (i.e. an environment similar to a non-aqueous solvent), seems to support the contention that the $H\phi O$ effect is the most important factor in protein folding. Unfortunately, the conclusion derived from these two facts is unjustified. First, the data on Gibbs energy of transfer are relevant to hydrophobic molecules, not to hydrophobic groups attached to proteins. Second, the fact that the majority of the $H\phi O$ groups are found in the interior of protein does not mean that the $H\phi O$ effect is the "driving force" for the folding process. See Sect. 1.4 and Ben-Naim [3, 4] for details.

A. Ben-Naim, *Solvent-Induced Interactions and Forces in Protein Folding*, Physical Chemistry in Action, https://doi.org/10.1007/978-3-031-31871-9_6

Once the new $H\phi I$ effects (including the old intramolecular HBs) were discovered the mystery associated with the PFP was largely evaporated. The $H\phi I$ effects could easily explain not only the stability of the 3D structure of proteins but other mysteries in biochemistry such as the self-assembly of protein. Furthermore, the discovery of strong $H\phi I$ forces (which were never studied before) could also explain the *speed* of folding; a problem sometimes referred to as the Levinthal Paradox.

In the next sections of this chapter, we shall show how the various $H\phi I$ effects (both interactions and forces) can explain most of the mysteries in protein chemistry.

6.1 The Problem of Stability of the 3D Structures of Proteins

In Chap. 1, we explained how the protein folding problem (PFP) became the main motivation for studying SII. We saw that the PFP consists of essentially three different problems. In this section, we discuss the problem associated with the stability of the 3D structures of proteins.

Figure 6.1 shows a schematic process of the folding of a protein. It is known that in order to *function* a protein must have a very precise 3D structure. It is also known that right after the protein is produced on the ribosome, it folds spontaneously into a precise 3D structure.

This process is sometimes "aided" by other proteins known as chaperones. However, since Anfinsen's works, it is believed that most of the proteins fold spontaneously in aqueous solutions. The main question is: Why?

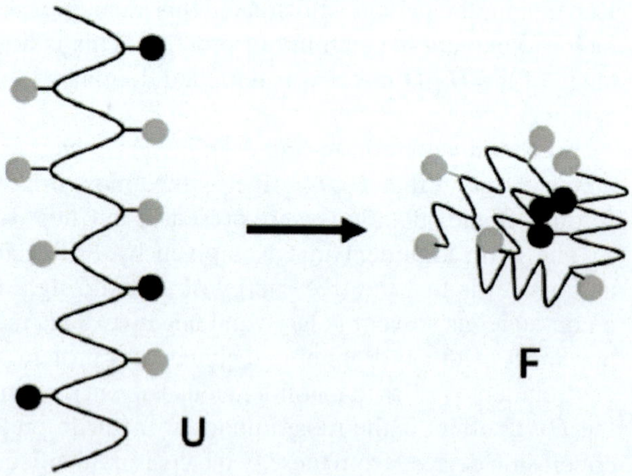

Fig. 6.1 Schematic process of folding of a protein. The unfolded form U is transformed into the folded form F

We write the chemical reaction of folding as:

$$\boldsymbol{U} \rightleftharpoons \boldsymbol{F} \tag{6.1}$$

where $\boldsymbol{U}$ and $\boldsymbol{F}$ denote the unfolded and folded forms, respectively. In order to get a significant concentration of the folded form, the standard Gibbs energy of this reaction must be large and negative. There are several ways of defining the *standard* Gibbs energy of a reaction. Nevertheless, in any definition of this quantity, it is always possible to write it as:

$$\Delta G^{\circ} = \Delta H^{\circ} - T\Delta S^{\circ} \tag{6.2}$$

In Eq. (6.2) ΔH° and ΔS° are the standard enthalpy and entropy change associated with the process of folding. The question we pose now is: How do we get a large negative value of ΔG°?

In the process of folding, Eq. (6.1), the unfolded molecule ($\boldsymbol{U}$) is transformed into the folded form ($\boldsymbol{F}$). One expects that the great reduction in the number of configurations in this process is associated with a negative entropy change (ΔS°). Therefore, in order to get a large negative ΔG°, we must have a very large enthalpy (or energy) change in the process.

As we discussed in Chap. 1 and in the introduction to this chapter, the earlier explanation of the origin of the required large and negative energy change, was based on the strength of the intramolecular HBs. Unfortunately, this source of energy change was believed to be "insignificant." Instead, it was suggested that the $H\phi O$ effect, here, associated with the transfer of non-polar groups from water into the interior of the protein, could contribute a large negative Gibbs energy change. This "explanation" of the folding "driving forces" dominated the literature (see Dill [2]). Unfortunately, it was found that this $H\phi O$ effect is quite weak (see discussion of the conditional solvation, Sect. 4.5). Therefore, the $H\phi O$ effect *cannot* explain the stability of the structure of the folded protein.

In the late 1980s and early 1990s a few, new $H\phi I$ effects were discovered (including direct intramolecular HBs). These were found to provide a large and negative "driving force" (here, in the thermodynamic sense) for the folding process. In my opinion, the discovery of these $H\phi I$ solved the problem of the stability of proteins, or equivalently explained the source of the large and negative standard Gibbs energy for the folding process.

Finally, it should be added that some proteins are crystallized with a few water molecules which are part of the 3D structure of the protein. In this case, it is likely that these water molecules form 3 or 4 HBs with the hydrophilic groups of the protein.

6.2 The "Speed" of the Folding Process

As we discussed in section…the so-called protein folding problem consists of three different problems. This problem appears in the list of 125 unsolved ("What don't we know?") questions in science, published in Science (Kennedy and Norman [5]):

> Can we predict how proteins will fold? Out of a near infinitude of possible ways to fold, a protein picks one in just tens of microseconds. The same task takes 30 years of computer time.

Clearly, these are two distinct problems; one, is the question about the "prediction" of the 3D structure of a protein, given its sequence of amino acids. The second involves the "speed" of the folding process. In this section, we shall discuss only the second question, which is associated with the so-called "Levinthal Paradox."

Levinthal [6, 7] wrote:

> (a) Let us ask ourselves how proteins fold to give such a unique structure. By going to a state of the lowest free energy? Most people would say yes and indeed, this is a very logical assumption. On the other hand, let us consider the possibility that it isn't so.
>
> (b) How accurately must we know the bond angles to be able to estimate these energies? Even if we knew these angles to better than a tenth of a radian, there would be 10,300 possible configurations in our theoretical protein. In nature, proteins apparently do not sample all of these possible configurations since they fold in a few seconds, and even postulating a minimum time for going from one conformation to another, the proteins would have time to try the order of 108 different conformations at most before reaching the final state.
>
> (c) We feel that protein folding is speeded and guided by the rapid formation of local interactions which then determine the further folding of the peptide. This suggests local amino acid sequences which form stable interactions and serve as nucleation points in the folding process.
>
> (d) Then, is the final configuration necessarily the one of the lowest free energy? We do not feel that it has to be. It obviously must be a metastable state which is in a sufficiently deep energy well to survive possible perturbations in a biological system. If it is the lowest energy state, we feel it must be the result of biological evolution, i.e. the first deep metastable trough reached during evolution happened to be the lowest energy state. You may then ask the question, "Is it a unique folding necessary for any random 150-amino acid sequence?" and I would answer, Probably not.

In this quotation, all three problems of the PFP appear. Parts (a) and (b) led people to think that Levinthal had found a "paradox" and suggested ways of how to solve this "Levinthal's Paradox." (See, for example, Zwanzig et al. [8], and Ben-Naim [3]). In Ben-Naim [3], it is shown that Levinthal did not discuss any "paradox" and the various "solutions" suggested only added to the confusion about the PFP, rather solved any paradox. If you read carefully Levinthal's article you will find several confusions between "energy" and "free energy," and also between "interactions" and "forces." But you will find no paradoxes.

As we have discussed in several places in this book, intramolecular *interactions* are important for explaining the *stability* of the final 3D structure of a protein. On the other hand, the question of the "speed" and "guidance" of the folding process is not determined by the interactions, but by the *forces* exerted on the various groups around the protein.

To understand the relationship between the *forces* and the *speed* of folding consider first the simplest case of the two particles, A and B at fixed positions in a solvent, say water, at some temperature T and pressure P.

Suppose that we release particle B and we observe that it will move towards A at a very high speed. How can we explain this "high speed?" The answer is that A exerts a strong force on B. If these two particles were in a vacuum, Fig. 6.2, the motion of B would have been fast and along a straight line toward A. In liquids, as we have discussed in Sect. 3.3, the force on B has two compartments: A direct force exerted by A (the same force as in a vacuum), and a solvent-induced force which is an average force exerted by the solvent molecules.

If the total force on B is not very strong, then we expect that B will fluctuate above and below the straight line toward A. In this case expect that the motion of B will be somewhat random with an overall preferential direction towards A, as shown in Fig. 6.3.

If the total force on B is very strong, then we expect that B will move fast and along nearly a straight line towards A. the fluctuations will be very small, as shown in Fig. 6.4.

As we explained in Chap. 3, the "interaction" between A and B is determined by the PMF at any distance R. The force is determined by the gradient of the PMF with respect to $\boldsymbol{R}_B$. The strength of the force could change with the distance; it could be

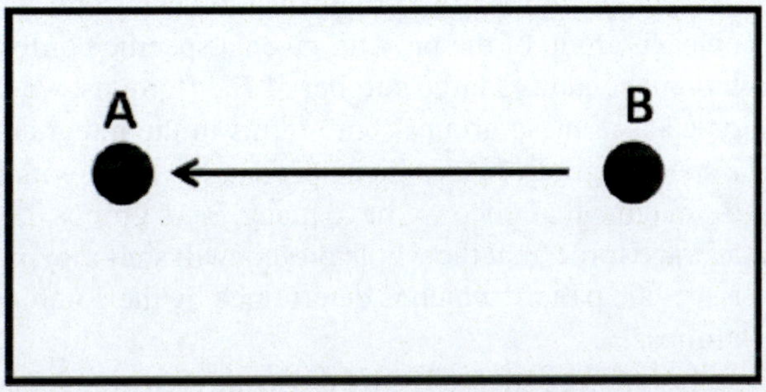

Fig. 6.2 Two balls in a vacuum. The lager the force acting on B the speedier the motion

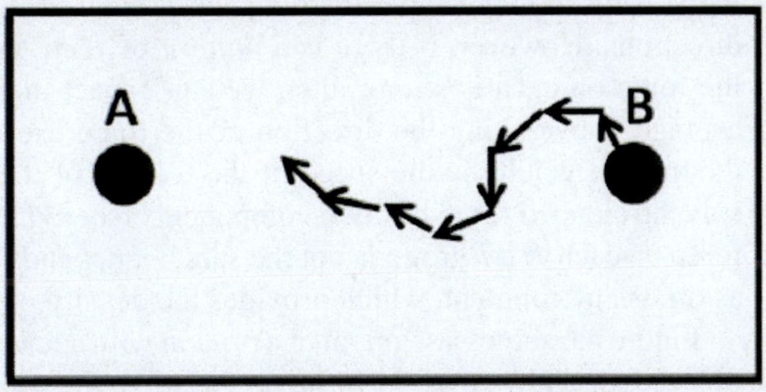

Fig. 6.3 Two balls in a solvent. The Force on B consists of two parts; a direct force exerted by A and an indirect part due to the presence of the solvent. The latter force will cause a random motion of B

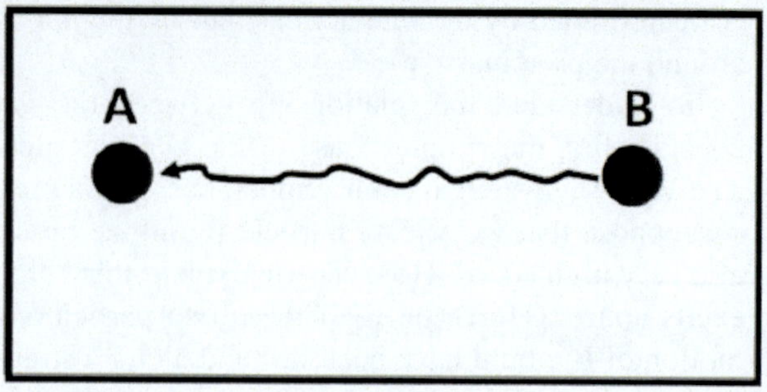

Fig. 6.4 Two balls in a solvent. When the direct force exerted by A is very large compared with the indirect force, the motion will be almost directly towards A

either attractive or repulsive at different distances. This force is an *average* force, not the instantaneous force operating on the particles.

Thus, we can conclude for the two particles in a solvent, if there is a strong force on say, B (including the direct and the indirect components), we could expect that the motion will be fast in the direction of the force. This direction and speed will be maintained if the force continues to be stronger in the successive sequence of times.

Before we move to discuss protein-folding, suppose that someone, call her Linda, never heard about Intermolecular forces. She assumes that once B is released, it will start moving "at random." A quick calculation might lead Linda to an estimate that the time to reach particle A would be, say one year. In reality, the time to reach A is about a second. Would you conclude that there is a paradox in the motion of the particles? Would you call that Linda's Paradox?

The same is true for a protein in a solvent. Here, we can ask about the average force on each group of the protein, given a specific conformation of the entire protein. A protein contains a large number of $H\phi I$ groups, we can expect strong forces exerted on each of these groups (both terms in the integrand of Eq. (3.38) Strong force on a specific group causes high speed and a preferential direction of motion. It is easy to imagine that once we have many $H\phi I$ groups, on each of these there is a strong average force, that the whole protein will start moving at with high speed, and follow a specific pattern which is determined by the combination of all the forces on all the groups.

It should be stressed as we did in the case of two-particles-in-a-solvent. The overall folding pattern will not be unique. There will always be random forces exerted by the solvent molecules bombarding the protein at different points and in different directions. However, if there is a *pattern* of $H\phi I$ groups along the protein, and if the force on each is strong, then we can expect that each of these groups will, on average, move along the direction of the force exerted on it. The larger the force, the highest would be the speed of the folding of the entire protein. Remember the solvent-induced force has two components [see Eq. (3.38)]. This means that the net force on each $H\phi I$ group is not the same; it depends both on the group itself as well as on its environment, which provides the density of water molecules, Fig. 6.5.

Figure 6.6 shows a segment of a protein with a few arrows indicating the directions of forces on a few groups. The larger the arrow the stronger the force. When the forces are very strong, it will be easy to "predict" how this segment will fold. It is clear that

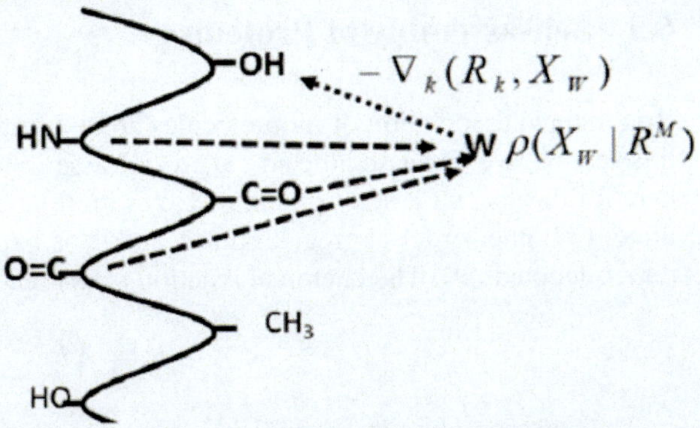

Fig. 6.5 The *indirect*, or the solvent-induced force on the OH group has two factors, see Eq. (3.38). A *direct* force due to a water molecule W (dotted arrow), and a local density of water molecules, which is due to other hydrophilic groups in the neighborhood (dashed arrows)

for any given pattern of hydrophilic groups, there will be a given pattern of forces. These forces will be changed as the configuration of the protein changes. However, it is clear that if the forces are strong enough, the pathway for the folding will be almost unique. And the stronger the forces the speedier would be the folding process.

In my view, the very fact that we know that there are strong $H\phi I$ forces, on many of the $H\phi I$ groups on a protein, already answers Levinthal's question about the speed of folding. Of course, for each specific protein there is a specific pattern of forces, and therefore a specific pathway of the folding.

Originally, Levinthal [6], estimated that if the process of protein folding went through a kind of a random walk, then it would have taken eons to reach the final structure. The fact is that the folding takes a few minutes only. This means that the assumption of random folding is incorrect. Obviously, a wrong prediction based on an incorrect assumption does not constitute a "paradox."

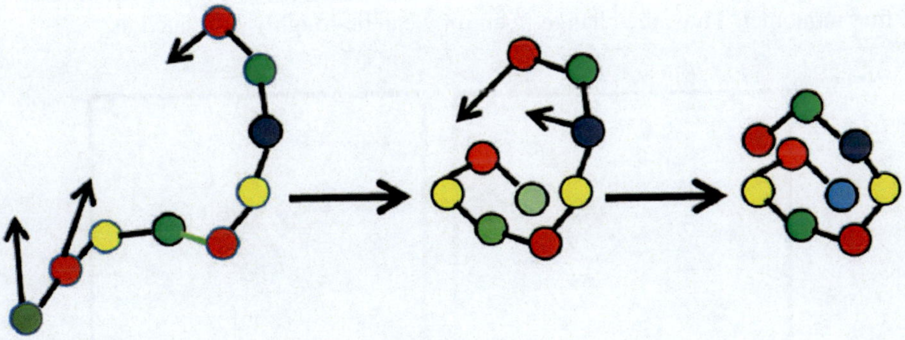

Fig. 6.6 A short segment of a protein. The arrows indicate the force on each group. When the forces are strong, the motion of the groups, along the protein would be determined by the pattern of those forces

6.3 Self-assembly of Proteins

Binding and association of biomolecules are ubiquitous in biochemical systems. In this section, we discussed the so-called self-assembly of proteins. This term is used here for the special case of binding two (or more) globular proteins to form a stable dimer or higher-order aggregate. Let A and B be two spherical proteins that form a dimer denoted AB. The chemical reaction is written as:

$$A + B \rightleftharpoons AB \tag{6.3}$$

The corresponding standard Gibbs energy change may be written as:

$$\Delta G^{\circ} = \Delta H^{\circ} - T\Delta S^{\circ} \tag{6.4}$$

In order to get a substantial amount of the dimer AB, we need to have a large and negative ΔG°. The problem (and the mystery) of the self-assembly of proteins is similar to the problem of protein folding. In the process of dimerization, there is always a large negative change in entropy due to the loss of the translational entropy; initially, both A and B can independently access the entire volume of the system V. Once they form a dimer, there is a reduction in the total accessible volume. In this case, one particle, say, A has access to the entire volume V, but for each location of A, B can access only a very small volume v around A, see Fig. 6.7.

This change in entropy may be interpreted in terms of Shannon's measure of information. Initially, we lack locational information on both A and B in the entire volume V. Once a dimer is formed we lack information on the location of *one* monomer (either A or B) in the entire volume V, and another smaller lack of conditional information on the location of the second monomer, given the location of the first monomer. Thus, the change in entropy can be roughly estimated as:

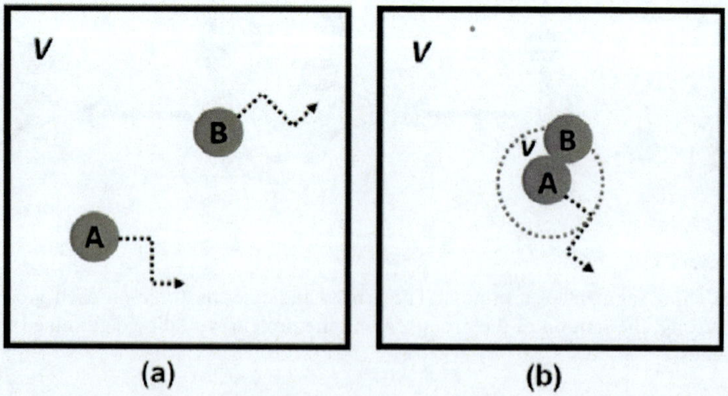

Fig. 6.7 Two spherical particles in a box of volume V. **a** Each particle can access the entire volume of the box independently. **b** When A and B form a dimer, then for each location of say, A, in the entire volume V, the second particle B, can access only a small volume v about the center of A

$$\Delta S^{\circ} \approx -k \ln \frac{V}{v} \tag{6.5}$$

In Eq. (6.5), V is the macroscopic volume of the system, and v is a small volume of the order of the *molecular* volume of the monomer.

Clearly, a large and negative value of ΔS° will contribute a large positive value to ΔG°. The question arises; what source of energy may be large and negative to compensate for this large and positive contribution due to the entropy change?

To understand the nature of this problem, consider the case of two argon atoms. We could write the same chemical reaction of dimerization as in Eq. (6.3) for this case. The change in entropy ΔS° will be large and negative, as explained above. Therefore, $-T\Delta S^{\circ}$ will be large and positive. Since we do not expect any large and negative interaction energy between the two argon atoms, we expect that ΔH° (or ΔE°) would be negligibly small compared to $-T\Delta S^{\circ}$. Therefore, we expect that ΔG° would be large and positive for argon. This means that the probability of finding a "dimer" of the two argon atoms is almost zero.

Next, consider the case of two hydrogen atoms. Here again, $-T\Delta S^{\circ}$ would be large and positive due to the loss of the translational entropy of one atom in forming the "dimer" (the hydrogen molecule H_2). However, unlike the case of two argon atoms, here, the two hydrogen atoms can form a strong chemical bond. Hence, ΔH° (or ΔE°) would be large and negative. This is enough to over-compensate for the positive value of $-T\Delta S^{\circ}$. This means a large and negative ΔG°, or equivalently a large concentration of the "dimer" H_2.

Going back to the dimerization of two globular proteins, the question is: What is the factor that will make ΔG° negative in spite of the loss of the translational entropy of one monomer?

There had been several attempts to estimate the thermodynamic quantities of association of specific proteins, such as insulin, see Steinberg and Scheraga [9], or to estimate specific contributions to ΔG° (see Makhatadze and Privalov [10], Amzel [11] and Ben-Naim [12]). An excellent review of this topic was published by Yu et al. [13].

In order to answer this question, one needs to calculate the ΔG° for the dimerization of two globular protein molecules. This calculation was undertaken in Ben-Naim [14]. First, we write the standard Gibbs energy of dimerization as:

$$\Delta G^{\circ} = \Delta G(TR) + \Delta U + \delta G \tag{6.6}$$

The first term $\Delta G(TR)$ is the change in the Gibbs energy due to the change in the translational and rotational (TR) degrees of freedom on the process of dimerization. In general, this term should include all the changes in the internal degrees of freedom in the process. However, in our calculation, we neglected all the changes except for the translational and rotational degrees of freedom. As expected, the quantity $\Delta G(TR)$ was found to be large and positive and a monotonically increasing function of the radius of the protein, Fig. 6.8. This means that the probability of finding a "dimer" would be negligibly small.

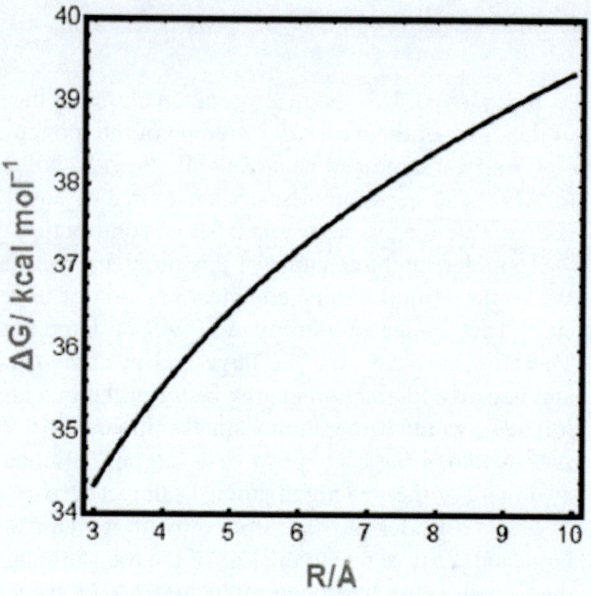

Fig. 6.8 The translational and rotational contribution to the standard Gibbs energy of dimerization $\Delta G(TR)$, as a function of the radius of the protein, at 298.15 K

The next step is to add the *direct* interaction between the two monomers at contact which we denoted by ΔU in Eq. (6.6). Figure 6.9 shows all possible *direct* interactions between the two globular proteins. All these direct interactions are essentially short-range. Since the region of contact between the two spheres is quite small we cannot expect a large contribution due to direct interaction between groups on the different monomers.

Next, we turn to SII between $H\phi O$ and $H\phi I$ groups on the surface of the protein. Figure 6.10, shows two pair of $H\phi O$ say, methyl groups which can interact through the solvent. We do not expect a large contribution from such SII; first, because there

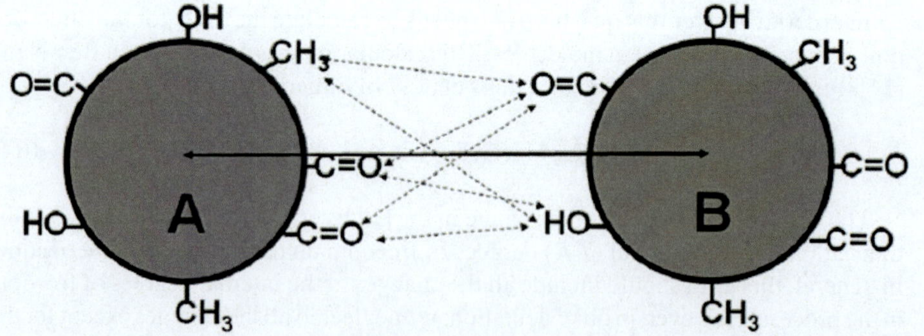

Fig. 6.9 The total direct interaction between two proteins. Only those pairs which are close enough are shown

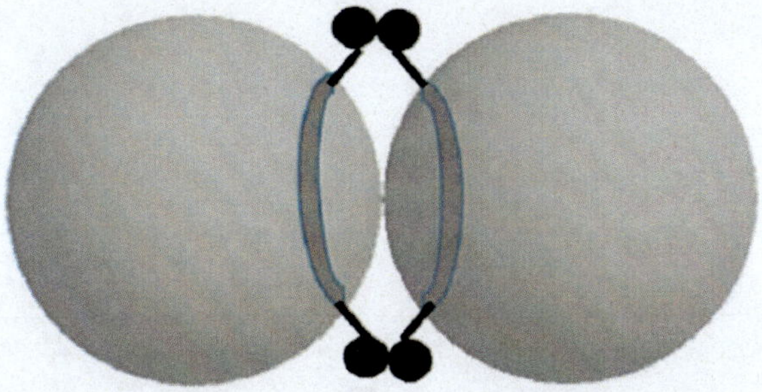

Fig. 6.10 Two pairs of hydrophobic groups (darker spheres) on the surface of the two spherical proteins can interact through the solvent. Such pairs of groups must be on a small strip (darker gray areas) on the surface of the two proteins NOTE THE FIGURE IS TOO BIG

are not too many $H\phi O$ Groups on the surface of the protein. Secondly, because these interactions are quite weak.

Thus, the largest possible interaction which would contribute large and negative ΔU° would be due to HBs between $H\phi I$ groups on one protein with $H\phi I$ on the other. It was estimated that any such *direct* interaction could not be large enough to overcome the entropy change in the association process.

Figure 6.11 shows two pairs of $H\phi I$ groups on the surface of the protein which can be bridged by a water molecule. Each such pair of $H\phi I$ groups should be at a right distance (and orientation) to form such a HBB, therefore they must be on a spherical strip on the surface of the protein, shown in darker gray color in Fig. 6.11. We can expect that these solvent-induced interactions could provide such a negative contribution to the total standard Gibbs energy change. Recall that there are many $H\phi I$ groups on the surface of the protein and each HBB, would contribute significantly large SII.

In 2006, I estimated the relative contribution of the $H\phi O$ and $H\phi I$ pair interaction between $H\phi O$ and $H\phi I$ groups on the two monomers, respectively. The details of the calculations are lengthy and will not be reproduced here (see Ben-Naim [12, 14]).

We summarize here only the conclusion reached in 2006 regarding the main factor that provides the largest contribution to the negative value of ΔG°.

We assume that the surface of the protein contains both $H\phi O$ (e.g. methyl group) and $H\phi I$ groups (e.g. C = O, or OH, NH) and these are distributed uniformly over the surface. Let $x(H\phi O)$ and $x(H\phi I)$ be the mole fractions of $H\phi O$ and $H\phi I$ groups, respectively.

$$x(H\phi O) + x(H\phi I) = 1 \tag{6.7}$$

When $x(H\phi O)$ is zero, all the groups on the surface of the protein are $H\phi I$.

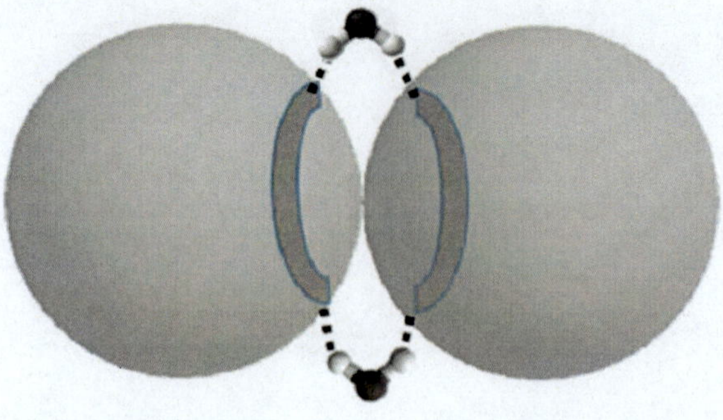

Fig. 6.11 Two pairs of hydrophilic groups on the surface of the two spherical proteins can form HBB by a water molecule such pairs of groups must be on a small strip (darker gray areas) on the surface of the two proteins NOTE THE FIGURE IS TOO BIG

We focus on *pair* of $H\phi O$ and $H\phi I$ interaction and neglect interactions between larger numbers of such groups. In this case, the SII can come only from functional groups which are in a narrow strip of two proteins, as shown in Figs. 6.10 and 6.11.

Figure 6.12 shows the results of the result of the calculation of the Gibbs energy change for the dimerization of two spherical protein, for three values of the mole fraction $x(H\phi I)$, 0, 0.5 and 1. When all the groups on the surface are $H\phi O$, we found that ΔG° is positive and will not be strongly dependent on the radius of the globular protein. When $x(H\phi I) = 0.5$, the Gibbs energy change is small and positive for small proteins and becomes large and negative as the radius of the protein increases. When $x(H\phi I) = 1$, we find a very strong dependence on the radius, and the values of ΔG° become large and negative, meaning a large probability of formation of "dimers" despite the loss of the translational entropy of one monomer.

Clearly, this kind of $H\phi I$ interaction may be operative for non-spherical proteins and also for association between more than two monomers.

As in the case of the protein folding problem (PFP), the mystery of self-assembly of protein appeared also in the list of 125 unsolved ("What don't we know?") questions in science, published in Science (Kennedy and Norman [5]):

How do proteins find their partners?

Protein–protein interactions are at the heart of life. To understand how partners come together in precise

orientations in seconds, researchers need to know more

about the cell's biochemistry and structural organization.

As we noted in connection with the protein folding problem, we believe that the discovery of $H\phi I$ interactions effectively removed the self-assembly problem from the list of "unknowns in science."

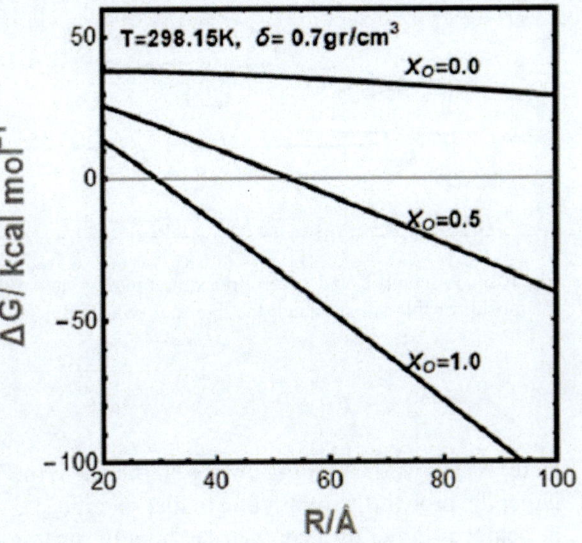

Fig. 6.12 Values of the Gibbs energy changes for the formation of a dimer between two globular proteins, for different mole fractions of the hydrophilic groups (x_0) on the surface of the protein, as a function of the radius of the protein

6.4 Solvent-Induced Effect in Molecular Recognition

In the previous section, we discuss the process of binding two, or more protein molecules to form stable "dimer," or higher-order aggregates. We saw that the most likely factors that provide the "thermodynamic forces" for binding the two monomers are the $H\phi I$ interactions. In this section, we discuss another process that might be affected by $H\phi I$ interactions. This is the process of binding a relatively small molecule (a ligand) to a larger macromolecule. Examples are the binding of oxygen to hemoglobin, binding a substrate to an enzyme, or binding protein to a specific site on the DNA.

The main question is the following: A small molecule (ligand) approaching the surface of a larger macromolecule is known to bind preferentially to a specific site. How does this ligand "choose" the correct site on which to bind? This question is known as the problem of molecular recognition (MR).

The problem of MR is of interest not only in understanding the driving force for such highly specific processes, but is of utmost importance in some practical problems such as *drug design*.

Until quite recently the dominating theory underlying the problem of molecular recognition was the so-called "lock and key" model, or lock-and-key mechanism proposed over a hundred years ago by Fischer (1897). In fact, almost all drug-design methodologies have been based on the lock-and-key model.

The lock-and-key model is based on the idea that when a ligand approaches a surface of a large macromolecule, it will preferentially bind to that site at which the *direct* binding energy is the strongest.

Figure 6.13 shows a ligand L approaching a large polymer P. The criterion for choosing a specific binding site is:

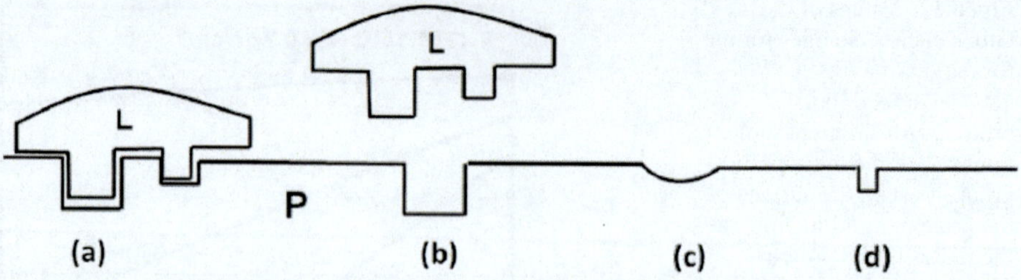

Fig. 6.13 A ligand L and several possible binding sites, (a) to (d), on a polymer P. In a vacuum, the most probable site at which the ligand L will bind is (**a**)

$$\min_i \Delta U(i) \tag{6.8}$$

Here, $\Delta U(i)$ is the binding energy to the site *i*, the larger (and negative) $\Delta U(i)$, the larger the probability of binding to that specific site. The minimal value of $\Delta U(i)$ can be achieved either by a geometrical fit or by a pattern of matching of some functional groups (such as $H\phi I$ group that can form direct HBs).

Clearly, the geometrical fit underlying the lock-and-key model is equivalent to maximizing the number of van-der-Waals interactions between groups on *L* and groups on *P*. Figure 6.13 shows a few binding sites for a ligand *L*. Clearly, according to the lock-and-key model, site (a) will be the preferable binding site. This is so because the *direct* interaction between *L* and *P* at this site is the largest.

Indeed, if *L* and *P* are in an ideal gas phase (or in a vacuum) than the probability ratio for binding on two sites, *i* and *j*, is related to the difference in the binding energies, i.e.

$$\frac{\Pr(i)}{\Pr(j)} = \exp[-(\Delta U(i) - \Delta U(j))/kT] \tag{6.9}$$

Hence, it follows that:

$$\max \Pr(i) = \max\{\exp[-\Delta U(i)/kT]\} \tag{6.10}$$

In Eq. (6.10) the maximum is over the index *i*. The situation is quite different when the binding occurs in an aqueous solution. In this case, the criterion for strongest binding, or highest probability of binding is:

$$\max \Pr(i) = \max\{\exp[-\Delta G(i)/kT]\} \tag{6.11}$$

where $\Delta G(i)$ is the Gibbs energy change for the process of binding (for a specific choice of standard states which will not concern us here).

Most people who work on drug design continue to use the lock-and-key model even after recognizing that the most likely site for binding is determined by the Gibbs energy criterion, Eq. (6.11), and not the binding energy criterion, Eq. (6.10).

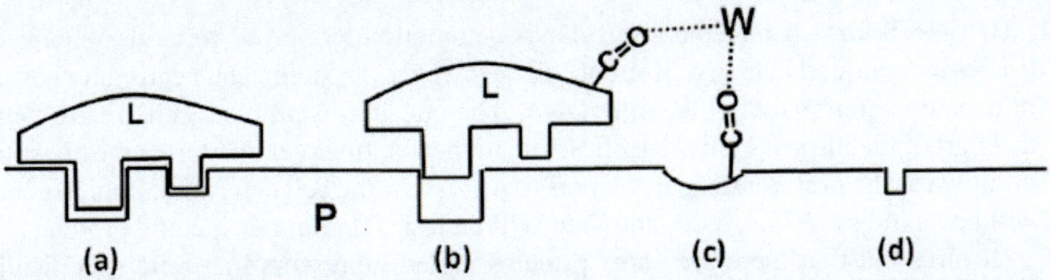

Fig. 6.14 A ligand L and several possible binding sites on a polymer P. In water, the most probable site at which the ligand L will bind is (**b**)

The rational for this is that people believe that if the ligand fits best to site i in a vacuum, it is likely that it also fits best to the same site when the binding occurs in a solvent.

This is unjustified, especially when the solvent is water, and when both the ligand L and the polymer P have $H\phi I$ groups which can be bridged by water molecules.

Consider the ligand L as shown in Fig. 6.14 and several possible sites on the surface of P. When the whole system of L and P is in an ideal gas phase, the ligand will prefer to bind to the site (a). The ratio of the probabilities in this case, is:

$$r^{(g)} = \frac{\Pr(a)}{\Pr(b)} = \exp[-\Delta U(a)/kT + \Delta U(b)/kT] > 0 \tag{6.12}$$

Clearly, in this case the direct binding interaction energy to site (a) is stronger than to site (b). Note also that the $H\phi I$ groups shown on L and P are assumed to be far apart and therefore these two groups do not contribute to the direct binding energy between L and the site on P.

When the same system is in water, the two $H\phi I$ groups can interact by forming a HB bridge as shown in Fig. 6.14. In this case, we expect that the SII between the two $H\phi I$ groups will dominate the binding Gibbs energy to the site (a). Equivalently the ratio of probabilities:

$$r^{(w)} = \frac{\Pr(a)}{\Pr(b)} = \exp[-\Delta G(a)/kT + \Delta G(b)/kT] < 1 \tag{6.13}$$

Thus, the preference for binding to the site (b) is due to the $H\phi I$ interaction, and the geometrical fit (i.e. the lock-and-key-model) becomes irrelevant to the determination of the preferred binding sites.

We believe that such a solvent-induced interaction might be important in drug design, especially for drugs that have a large number of $H\phi I$ groups, which bind to surfaces having a specific pattern of $H\phi I$ groups. For some specific examples of binding sites on proteins, see Wang and Ben-Naim [15].

Another possible involvement of strong $H\phi I$ interactions is in the case of binding of certain proteins to a specific site on the DNA. This is one of the most important processes in molecular biology.

It is well known that the entire DNA—sometimes referred to as the "book of life"—is contained in many different cells. However, a specific cell needs only partial information contained in the entire DNA. The question is how the cell "reads" only that part of the information on the DNA that it needs. In molecular terms the question is known as the preferential gene expression, i.e. how the cell determines which genes will be "expressed," i.e. read, and later will be translated into a specific protein.

It turns out that there are some proteins called repressors that bind specifically on certain sites along the DNA (referred to as the operator), and by doing so they determine whether the specific gene that follows will or will not be expressed. The details of the mechanism of this "genetic switch" were developed by Ptashne [16]. It is a fascinating problem involving the "cooperativity" of binding which is discussed in detail in Ben-Naim [17].

We are mentioning this problem in the context of the more general problem of molecular recognition. In my opinion, there is no doubt that the repressor protein "recognizes" the specific binding site on the DNA by means of $H\phi I$ interactions.

This contention stems from the fact that the surface of the DNA, unlike that of a protein looks quite uniform from the structural point of view. A protein approaching a DNA is unlikely to find a site that is a perfect fit for its geometry. Most of the surface of the DNA consists of $H\phi I$ groups almost uniformly distributed on the DNA's surface.

Therefore, an approaching protein might find a specific pattern of $H\phi I$ groups on the surface of the DNA with which they can interact through $H\phi I$ interactions.

6.5 Solvation and Solubility of Globular Proteins

In the previous sections of this chapter, we discussed two "great" problems in biochemistry. These were and still are considered to be unsolved or "unknown" problems in Science (see Kennedy and Norman [5]). In my opinion, the discovery of the strong $H\phi I$ interactions effectively removed the mystery associated with both the protein-folding and the self-assembly of proteins from the list of "unknowns."

In this section we discuss a problem that does not feature in the list of "unknowns," and in fact, most people do not consider it as a problem at all.

If you ask a biochemist why globular proteins are soluble in water, the immediate response you will most likely get is that the globular proteins have a highly $H\phi I$ surface (recall that most of the $H\phi O$ groups are buried in the interior of the protein). Since these $H\phi I$ groups interact favorably with water (they are *hydrophilic*), we expect that the solvation Gibbs energy of such proteins will be large and negative, hence their solubility is expected to be large.

This answer is essentially correct. It seems that there is no mystery in the phenomenon of the large solubility of proteins in water.

During my sabbatical year at NIH, I heard from Harry Saroff about some unusual experiment observations he made about the solubility of proteins. He observed that sometimes the solubility of the protein would decrease by replacing one $H\phi O$ group with a $H\phi I$ group. Saroff could not explain those findings. He repeatedly told me that in his opinion the solubility of proteins is "no less of a mystery than either the protein folding or the self-association problems."

At that time, I did not see what the mystery associated with the solubility of the protein is. It was only after the discovery of the $H\phi I$ interactions in the 1990s that it became clear to me.

Before I discuss the problem of the high solubility of globular proteins in water, let me ask the reader a simpler question. Figure 6.15a shows two isomers, Fumaric and Maleic acids. They have the same chemical formula and the same number of $H\phi I$ groups exposed to the solvent. Can you guess which of these two molecules would be more soluble in water, and why?

I will provide the answers at the end of this section.

I will first discuss here the relationship between solvation Gibbs energy of a solute α and its solubility.

We assume that pure α is a solid at the temperature T, and pressure P at whose solubility we are interested in Fig. 6.16.

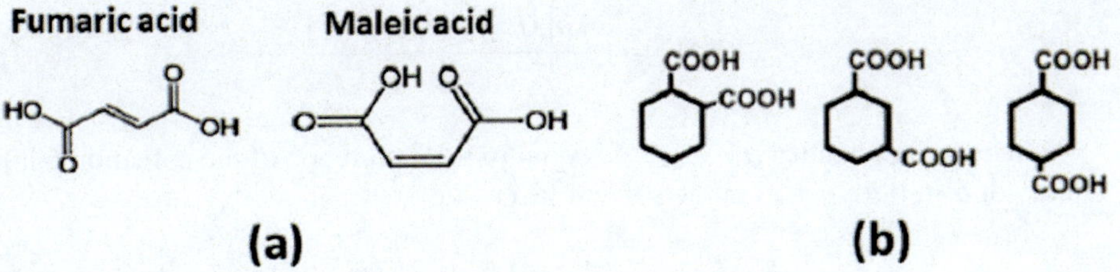

Fig. 6.15 **a** Two isomers of dicarboxylic acids, Fumaric and Maleic. **b** Three isomers of benzene dicarboxylic acids

Fig. 6.16 The solubility of α is the concentration of α at equilibrium between the pure phase α and the solution

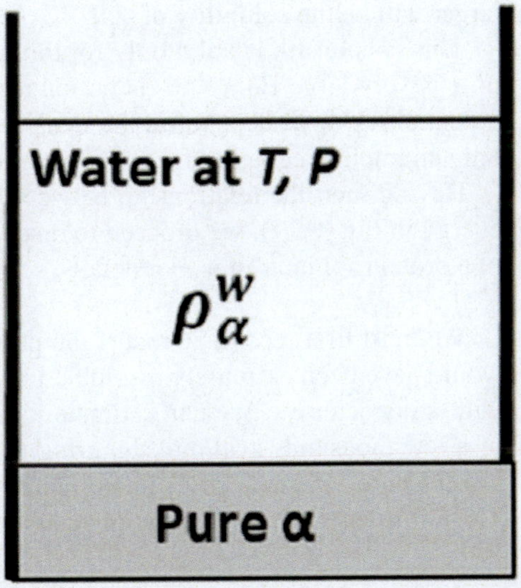

The solubility of α in the solvent w is defined as the concentration of α in w at equilibrium. The condition of chemical equilibrium is:

$$\mu_\alpha^p = \mu_\alpha^w \tag{6.14}$$

Here, μ_α^p is the chemical potential of α in the pure phase (p) and μ_α^w is the chemical potential of α in the solvent (w).

If we know that the solubility of α in w is very low, so that it forms a dilute-ideal solution, then we can write the condition of equilibrium as:

$$\mu_\alpha^p(P, T) = \mu_\alpha^{*w}(P, T) + kT \ln \rho_\alpha^w \Lambda_\alpha^3 \tag{6.15}$$

$\mu_\alpha^{*w}(P, T)$ is the pseudo-chemical potential of α in w (see Ben-Naim [14]). This quantity is a function of P and T, and does not depend on the density of the solute ρ_α^w. Λ_α^3 is the momentum partition function of α.

Clearly, at equilibrium the solubility ρ_α^w is determined by the equation:

$$\rho_\alpha^w(P, T) = \frac{\exp\left[\left(\mu_\alpha^p - \mu_\alpha^{*w}\right)/kT\right]}{\Lambda_\alpha^3} \tag{6.16}$$

Note that the quantity $\mu_\alpha^p - \mu_\alpha^{*w}$ is equal to the difference of the solvation Gibbs energy of α in pure p, and in the solvent w, i.e.

$$\begin{aligned} \mu_\alpha^p - \mu_\alpha^{*w} &= \mu_\alpha^p - \mu_\alpha^{*ig} - \mu_\alpha^{*w} + \mu_\alpha^{*ig} \\ &= \Delta G_\alpha^{*p} - \Delta G_\alpha^{*w} \end{aligned} \tag{6.17}$$

Clearly, the larger (in absolute magnitude) the solvation Gibbs energy ΔG_α^{*w}, the larger will be the solubility of α.

This conclusion is valid only for dilute solutions of α in w. When the solubility of α is very large Eq. (6.16) is no longer valid. In such case, the pseudo-chemical potential μ_α^{*w} will depend on the density of the solute, and therefore Eq. (6.16) is only an implicit equation for ρ_α^w (which now appears on both sides of the equation).

Having seen the relationship between the solubility ρ_α^w and the solvation Gibbs energy in Eq. (6.17), we proceed to discuss very briefly the solvation Gibbs energy of a protein solute α in w. For details, see Wang and Ben-Naim [18], and Ben-Naim [12].

We must first recognize that if the protein had no $H\phi I$ groups on its surface, it would have been extremely insoluble in water. Suppose we start with a hard sphere with a diameter σ_p. We can estimate its solvation Gibbs energy to be of the order of a few thousands kcal/mol. Ignoring the solvation Gibbs energy of such a HS in a solid pure HS, we can estimate that its solubility in water will almost be zero. The Gibbs energy of solvation of such a HS solute may be estimated by the scaled particle theory (SPT). See details in Ben-Naim [12]. It was estimated that for a HS of a diameter of about 60 − 80 A°, the values of ΔG_{α}^{*} are very large and positive.

Next, we can add the "soft" van-der-Waals interaction. By this we mean that the whole protein is viewed as a spherical aggregate of methane molecules. In this case the whole protein may be viewed as a giant $H\phi O$ molecule. A rough estimation shows that the Gibbs energy of solvation of such a molecule will still be very large and positive.

The most surprising finding was that if we add as many $H\phi I$ groups as in typical proteins but we distribute them randomly on the surface of the protein, we get, as expected, a negative contribution to the solvation Gibbs energy. However, even with all these $H\phi I$ groups on its surface, the solvation Gibbs energy is still positive. Not as large as the "giant $H\phi O$" molecule, but still large and positive, which means a very small solubility. Thus, we see that if we take a real globular protein with whatever number of $H\phi I$ groups on its surface, this in itself does not guarantee a large solubility. In other words, the common explanation of the high solubility of globular proteins, based on the fact that their surfaces contain many $H\phi I$ groups, is not enough.

The final calculation we did was to rearrange the distribution of $H\phi I$ groups on the surface in such a way that these groups can be bridged by water molecules. We found that for each pair of $H\phi I$ brought to a distance of about 4.5 Å, there was a decrease of the Gibbs energy of solvation by about 2.5 − 3 kcal mol^{-1} which corresponds to an increase of solubility by a factor of about 120.

Figure 6.17 shows two spherical proteins, on their surfaces we find the same number of $H\phi I$ groups. Calculations show that the one on the right-hand side will be about more soluble in water than the one on the left-hand side, by a factor of about 240.

Thus, we can conclude that the large number of $H\phi I$ groups on the surface of the protein is indeed responsible for the large solubility. However, we learned that it is not the mere *number* of $H\phi I$ which is important but the specific distribution of distances between pairs and triplets of such groups that converts a highly insoluble macromolecule into a highly soluble one.

Once we recognize the importance of the distance-distribution of the $H\phi I$ group, it will be easy to explain how replacing one $H\phi I$ group by a $H\phi O$ group can *increase* the solubility, and how replacing a $H\phi O$ group with a $H\phi I$ group can *decrease* the solubility. Some specific examples are shown in Ben-Naim [12]. Basically, the idea is that replacing a $H\phi O$ group by $H\phi I$ group, might destroy one possible HBB and therefore resulting in a decrease in solubility. On the other hand replacing a $H\phi I$ by a $H\phi O$ group might create a new possibility for a HBB, and hence increases the solubility.

Finally, regarding the question I asked at the beginning of this section, about the solubility of maleic and fumaric acids. Here are the values of the solubility of the two isomers in water (at 25 °C and atmospheric pressure):

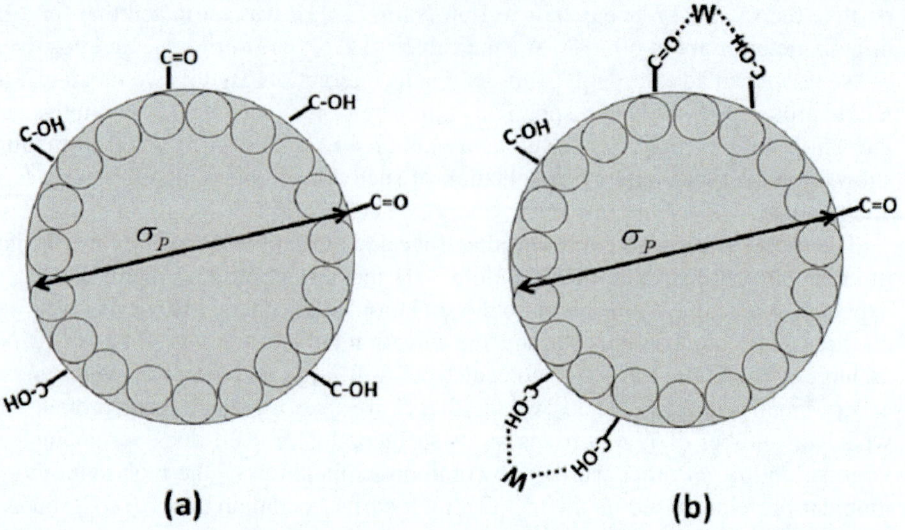

Fig. 6.17 Two spherical proteins. On both of their surfaces we find the same number of $H\phi I$ groups. However, in (**b**) the distribution of the $H\phi I$ groups is such that hydrogen-bond bridges are possible. No such bridges are possible in (**a**)

$$\begin{aligned}\text{Maleic acid} &\approx 78\ \text{gr}/100\ \text{ml}\\ \text{Fumaric acid} &\approx 0.6\ \text{gr}/100\ \text{ml}\end{aligned} \tag{6.18}$$

Thus, although these two molecules have the same number of $H\phi I$ groups, the solubility of maleic acid is larger than the solubility of fumaric acid by a factor of over 100. Why? As we noted above the solubility is determined by both the solvation Gibbs energy of a solute in water and in its pure solid state. In my opinion, it is unlikely that the Gibbs energy of solvation into the pure solute would be very different for these two isomers. Instead, the most likely explanation is the difference in the solvation Gibbs energy of the two isomers. Although these two isomers have the same number of $H\phi I$ groups, it is only in maleic acid that a water bridge can form connecting the two $H\phi I$ groups, Fig. 6.15a.

This interpretation is consistent with our estimate of the strength of the $H\phi I$ interaction. See Appendix B which leads to a factor of about 100 in solubility.

$$\begin{aligned}&\delta G(H\phi I\text{interaction in maleic acid})\\ &\approx -kT\ln\frac{78}{0.6}\approx -3\ \text{kcal mol}^{-1}\end{aligned} \tag{6.19}$$

A similar value was found for the indirect work (δG) for the transfer of a carboxylic group from position 4 to position 2 in benzene 1,2 di-carboxylic acid, Fig. 6.15b.

6.6 Cold Denaturation of Proteins

We start with a quotation from an excellent review by Privalov (1990), [19]:

> … disruption of the native structure upon heating, the heat denaturation of protein, appear to be an obvious effect. By the same argument, a decrease of temperature should only induce processes leading to increasing order.

Indeed, it is quite obvious that denaturation will occur at high temperatures. We write the standard Gibbs energy of folding as:

$$\Delta G^{\circ}(\boldsymbol{U} \rightarrow \boldsymbol{F}) = \Delta H^{\circ} - T\Delta S^{\circ} \tag{6.20}$$

ΔS° of folding is expected to be negative (not because of "ordering" but simply due to the larger number of possible conformations of the unfolded form $\boldsymbol{U}$). Therefore, assuming that the temperature dependence of both ΔH° and ΔS° is not very large, we can expect that as the temperature increases the term $T\Delta S^{\circ}$ will become more *negative*, hence $-T\Delta S^{\circ}$ will contribute *positively* to ΔG° eventually causing it to be positive. Hence, *folding* is unlikely, meaning unfolding (or denaturation) is likely.

Let us "rewind" the film and start from a high temperature at which all the protein is in the unfolded state $\boldsymbol{U}$. We now start cooling the system and we find that at some temperature the protein folds, i.e. x_F increases from zero to nearly one. The "driving force" for this process, the specificity, and the speed of the process are part of the protein folding problem (PFP). This is the famous experiment that was carried out by Anfinsen.

The new phenomenon referred to as Cold denaturation (CD) occurs when we further cool the system. Normally, further cooling will either cause an increase of x_F towards one, or if x_F already reached the limiting value of one, it will stay there.

Equivalently, we expect that the standard Gibbs energy $\Delta G^{\circ}(\boldsymbol{U} \rightarrow \boldsymbol{F})$ to be a monotonically increasing function of T, starting from negative values at low temperatures and eventually becoming positive at higher temperatures, i.e. we expect that:

$$-\Delta S^{\circ} = \frac{\partial \Delta G^{\circ}}{\partial T} > 0 \tag{6.21}$$

which is consistent with our expectation that $\Delta S^{\circ}(\boldsymbol{U} \rightarrow \boldsymbol{F})$ should be negative.

The occurrence of CD is unintuitive. It requires that below certain temperature proteins start to unfold, i.e. x_F decreases, or equivalently $\Delta G^{\circ}(\boldsymbol{U} \rightarrow \boldsymbol{F})$ starts to increase. This is quite a mystery.

Most of the conventional explanations are based on the idea that $H\phi O$ interactions become weaker at low temperatures, and since it is believed that $H\phi O$ effects are responsible for the folding of proteins, it follows that weakening of the $H\phi O$ effect will enhance unfolding. As we have seen in Chap. 4 and in Ben-Naim [3, 4] the $H\phi O$ effects (both solvation and interaction) are quite small and cannot explain the CD.

As in the case of the PFP, we now show that the combination of two kinds of $H\phi I$ effects can explain the phenomenon of CD. Recall that the $H\phi O$ effects (both solvation and interaction), become stronger as the temperature increases (this is true, at and about the range of temperature at which proteins exist in the folded form). The various $H\phi I$ effects have quite different temperature dependence. We focus here only on two pairwise $H\phi I$ effects; one, the direct formation of a HB between the two groups which we denote by $\delta G^{(2)}(HB)$, and second, the SII between two $H\phi I$ groups as a distance of about 4.5 Å which we denote by $\delta G^{(2)}(H\phi I)$. These are the analogs of the two final configurations of water molecules shown in Figs. 5.1 and 5.2.

As we have seen in Chap. 5, $\delta G^{(2)}(HB)$ is of the order of -2 kcal/mol, and $\delta G^{(2)}(H\phi I)$ is of the order -2.5 kcal/mol. Both of these effects are operative in stabilizing the 3D structure of the protein. However, it is clear that *direct* HBing are abundant in the ***F*** form (i.e. most of the $H\phi I$ groups in the interior of the protein are HBed). On the other hand, the indirect $H\phi I$ interactions exist in both the ***U*** and ***F*** forms. Whenever two $H\phi I$ groups are in the right configuration to form a water bridge, they will contribute to $\delta G^{(2)}(H\phi I)$.

From the scant experimental data that was available in 2006, it was found that $\delta G^{(2)}(HB)$ becomes *stronger* (i.e. it becomes more negative) as T increases, whereas $\delta G^{(2)}(H\phi I)$ becomes weaker (i.e. it becomes less negative) as T increases.

The estimated temperature dependence that was available at that times is shown in Fig. 6.18. It was later confirmed by simulated calculations as discussed in Chap. 5.

Here is the general qualitative explanation of the phenomenon of CD:

At relatively high temperatures the direct HBs are the main factor that stabilizes the 3D structure. Increasing the temperature makes this effect *stronger*. However, since temperature always "wins," the increase in temperature will always lead to *denaturation.* Thus, the strengthening of $\delta G^{(2)}(HB)$ cannot prevent denaturation. At most, it can "delay" its occurrence.

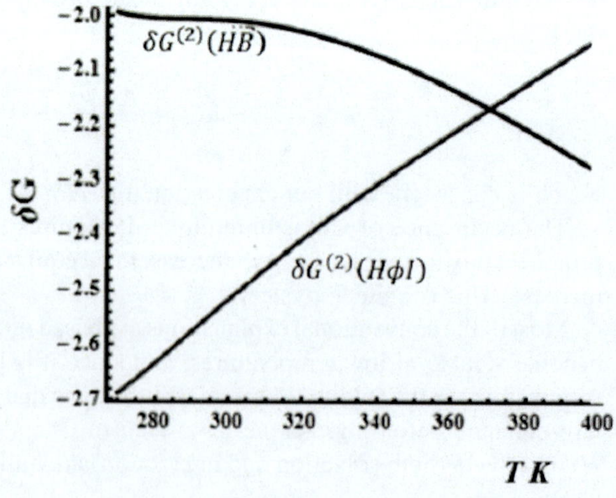

Fig. 6.18 The two different temperature dependence of two different hydrophilic effects, in kcal/mol, as estimated in Ben-Naim [14]

When we decrease the temperature, the effect of the indirect $H\phi I$ interaction increases. Since this $H\phi I$ effect works mainly to stabilize the $\boldsymbol{U}$ form (simply because there are more $H\phi I$ groups exposed to the solvent), we expect that denaturation will occur at a lower temperature.

Enough for qualitative discussion. Let us try to some calculations. In the following we will describe only the essential assumptions that were made in the calculations as well as provide some results. More details may be found in Ben-Naim [3, 14].

To simulate both heat and cold denaturation we assume the following solvent-induced effects:

1. Desolvation Gibbs energy of a $H\phi O$ group in the $\boldsymbol{U}$ form.
2. "Solvation" Gibbs energy of the desolvated $H\phi O$ group. This involves the van-der-Waals interaction of the $H\phi O$ group with its surroundings in the $\boldsymbol{F}$ form.
3. Desolvation Gibbs energy of $H\phi I$ group in the $\boldsymbol{U}$ form.
4. "Solvation" of the $H\phi I$ group in the $\boldsymbol{F}$ form. This involves mainly the intramolecular HB between the desolvated $H\phi I$ group and other $H\phi I$ groups in the $\boldsymbol{F}$ form.
5. Pairwise $H\phi O$ interaction in the $\boldsymbol{U}$ form.
6. Pairwise $H\phi I$ interaction in the $\boldsymbol{F}$ form.

A typical protein segment consists of three $H\phi O$ groups, and 12 $H\phi I$ groups. This is nearly the ratio of the two types of groups in real proteins. We assume that there are two intramolecular HBs, and three van-der-Waals interactions in the $\boldsymbol{F}$ form.

Thus, in an ideal gas phase the equilibrium constant for the reaction $\boldsymbol{F} \to \boldsymbol{U}$ is:

$$K^{ig} = \frac{q_F}{q_u} \tag{6.22}$$

where q_F and q_u are the internal partition functions of the $\boldsymbol{F}$ and $\boldsymbol{U}$ forms. These are:

$$q_F = \exp[-(2\varepsilon_{HB} + 3\varepsilon_{VDW})/kT] \tag{6.23}$$

$$q_F = N_C \tag{6.24}$$

ε_{HB} is the HB energy (about -6.5 kcal/mol). ε_{VDW} is about -0.5 kcal/mol, while N_C is the number of configurations of the U forms. This was chosen to be 10^{12} to account for the larger number of configurations of the $\boldsymbol{U}$ form.

Based on Eqs. (6.23) and (6.24), we can calculate the mole fraction of the F form from:

$$x_F = \frac{K^{ig}}{1 + K^{ig}} \tag{6.25}$$

As expected at very low temperatures x_F starts with the value of 1. When T increases x_F drops quite sharply to 0 at about $T \approx 260$ K. These results can be translated into Gibbs energy changes for the folding process:

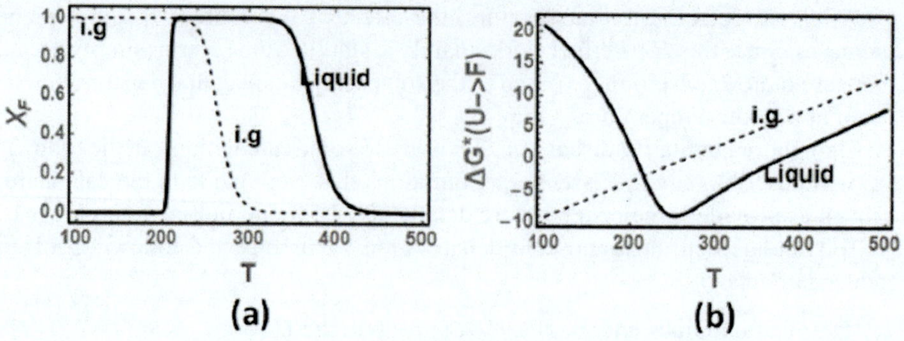

Fig. 6.19 **a** The mole fraction of the F form and **b** the Gibbs energy change as a function of temperature

$$\Delta G^{*ig} = \mu_F^{*ig} - \mu_u^{*ig} = -RT \ln K^{ig} \tag{6.26}$$

where μ_F^{*ig} and μ_u^{*ig} are the pseudo-chemical potentials of $\boldsymbol{F}$ and $\boldsymbol{U}$, respectively.

Again, as expected ΔG^{*ig} is a monotonically increasing function of temperature.

Next, we modify the equation for the chemical equilibrium, instead of Eq. (6.21) we can write the equilibrium constant for the same reaction in the liquid phase (w) as:

$$K^w = K^{ig} \exp\left[-\left(\Delta G_F^* - \Delta G_U^*\right)/kT\right] \tag{6.27}$$

where ΔG_F^* and ΔG_U^* are the solvation Gibbs energies of $\boldsymbol{F}$ and $\boldsymbol{U}$, respectively.

With additional information on the temperature dependence of the two $H\phi I$ effects, we can calculate both the mole fraction x_F^w and Gibbs energy change $\Delta G^{*w}(\text{U} \rightarrow \boldsymbol{F})$ as a function of the temperature Fig. 6.19.

Figure 6.19 shows that the Gibbs energy change for the folding process passes through a minimum and the x_F^w starts from 0 at low temperatures climb to 1, and remains at that value for a range of temperatures between ≈ 200 K and 380 K, then drops to 0 again as in the "normal" case of denaturation.

In Ben-Naim [3], on can find much more results on calculations of x_F^w and $\Delta G^{*w}(\text{U} \rightarrow \boldsymbol{F})$ for various values of the $H\phi I$ interactions. It was found that as the $H\phi I$ increases (in absolute magnitude) the range of temperatures in which the F form dominates, increases.

Thus, we see that with two different temperature dependences of the two $H\phi I$ effects we could reproduce the phenomenon of CD. For more details, the reader is referred to Ben-Naim [3, 14].

It should be noted that in order to obtain both heat and cold denaturation *without* $H\phi I$ interactions, one needs to take values of the $H\phi O$ solvation and $H\phi O$ interactions at least 15 times *larger* than the real experimental values.

6.7 Concluding Remarks

As we have noted several times throughout this book there are some real difficulties which are often referred to as "mysteries" in understanding the processes of protein folding and protein–protein association. It has long been recognized that the particular properties of water hold the keys to understanding these phenomena. However, the precise way in which water molecules are involved in these processes was long unknown. Therefore, it was not surprising to find these two problems in the list of "unknowns" in Science (2005), [5]. It should be added that the problem of the "structure of water" was also featured in the list.

Ever since Kauzmann suggested that the "$H\phi O$ bond" might be important in maintaining the stability of proteins, people grabbed the idea and accepted it as if offered a "solution" to the mysteries associated with both the folding and the self-assembly processes. This is quite unfortunate. First, because the $H\phi O$ effect was not well-defined, and certainly not understood, and second, when it was defined and its value estimated it was found that it was too weak to explain any of these processes.

What is more surprising is that people continued to "cling" to the $H\phi O$ effects as the "solution" for all the mysteries in biochemical processes even after it was shown that various $H\phi O$ effects are weak. Fortunately, we now know that the various $H\phi I$ effects are far stronger, and in principle could explain both the protein folding and the association of proteins.

In a review article titled "Energy Landscape and Solved Protein-Folding Problem," Wolynes [20], writes:

> Evolution solved the protein-folding problem. A major goal of bio-molecular science has been to understand how this was done.

Most of the article discusses "Funneled landscapes" and surveys the "contribution" of the "energy landscape theory" to our understanding of the protein folding problem.

In my opinion the "Energy landscape theory" does not exist, and certainly it does not contribute anything to the understanding of the PFP. I criticized this "theory" in Ben-Naim [3, 4]. Regarding the statement quoted above about evolution that solved the PFP, I would like to comment that in my opinion, this is a totally meaningless statement.

Evolution does not *face* problems, and evolution does it *solve* any problem. That is not the way evolution works.

During evolution, proteins were synthesized. Some proteins folded whereas others did not. Some of the folded proteins were useful and therefore survived for the simple reason that they had some advantages. Thus, it is meaningless to say that Evolution "solved" the PFP, and it is equally meaningless to say that we can "learn" as to how Evolution solved the PFP. In short, evolution does not proceed towards some pre-defined *target*, i.e. towards a protein that folds into a useful structure.

This kind of argument sometimes appears in popular science books. One can find statements about the DNA that was written in the "language of God." (See references in Ben-Naim [21]). It is known that the sequence of bases in DNA is translated into a sequence of amino acids in proteins, and the sequence of amino acids folds into a precise 3D structure. The last step is also referred to as the "second half" of the translation of the "information" contained in the DNA which was written in the "language of God." It follows that the folding into the precise (and useful) protein was also solved by God, using his own unique language. Extending these "logical" arguments perhaps we can learn from God himself as to how he solved the protein folding problem.

Even if one were to believe that God created all proteins in whichever language, it is far from clear that God himself was aware of the protein folding problem.

References

1. Kauzmann W (1959) Some factors in the interpretation of protein denaturation. Adv Protein Chem 14:1–63
2. Dill K (1990) Dominant forces in protein folding. Biochemistry 29:7133–7155
3. Ben-Naim A (2013) The protein folding problem and its solutions. World Scientific, Singapore
4. Ben-Naim A (2016) Myths and verities in folding theories. World Scientific, Singapore
5. Kennedy D, Norman C (2005) What don't we know? Science 309:5731–5775
6. Levinthal C (1968) Are there pathways for protein folding. J de Chimie Phys 65(1):44–45
7. Levinthal C, De Brunner JTP, Munck E (eds) (1969) Mossbauer spectroscopy in biological systems. In: Proceeding of a meeting held at Allerton House, Monticello, Illinois, Chicago, Illinois, University of Illinois Press, pp 22–24
8. Zwanzig R, Szabo A, Bagchi B (1992) Levinthal's paradox. Proc Natl Acad Sci USA 89:20–22
9. Steinberg IZ, Scheraga H (1963) J Biol Chem 238:172
10. Makhatadze GL, Privalov PL (1993) J Mol Biol 232:639
11. Amzel LM (1997) Proteins Struct Funct Genet 28:144
12. Ben-Naim A (2011) Molecular theory of water and aqueous solutions, part II: the role if water in protein folding, self-assembly and molecular recognition. World Scientific, Singapore
13. Yu YB, Privalov PL, Hodges RS (2001) Biophys. J 81:1632
14. Ben-Naim A (2006) Molecular theory of solutions. Oxford University Press, Oxford, UK
15. Wang H, Ben-Naim A (1996) J Med Chem 39:1531
16. Ptashne M (2004) A genetic switch, third edition: phage lambda revisited, 3rd edn. Cold Spring Harbor Laboratory Press, New York
17. Ben-Naim A (2001) Cooperativity and regulation in biochemical systems. Kluwer/Plenum Publ. New-York
18. Wang H, Ben-Naim A (1997) J Phys Chem B 101:1077
19. Privalov PL (1990) Cold denaturation of proteins. Biochem Mol Biol 25:281–305
20. Wolynes PG (2005) Energy landscapes and solved protein-folding problem. Phil Trans R Soc A 363:453–467
21. Ben-Naim A (2021) "Science of God", the greatest deception. Independent publisher, KDP, Amazon

Appendix A
The BN3D Model of Water and the Corresponding Pair Potential

In this appendix we describe very briefly the model of water which was deemed useful for both theoretical works on the outstanding properties of, and for estimating the strength of the $H\phi I$ interaction as well as the solvation Gibbs energy of water in water (see Appendix B). This model was first published in Ben-Naim [1]. Later, similar models for 1D water and 2D water molecules were published. (For details, see Ben-Naim [1, 2]). The three-dimensional model was only recently denominated as BN3D, the reasons for which were explained in Ben-Naim [2], Chap. 2.

A single water molecule is modeled as composed of two parts; a LJ particle with parameters as for neon atoms. This part is responsible mainly for the hard-repulsive part of the interaction between two molecules. The second part, is the HBing capability of the molecule, along four unit-vectors pointing to the vertices of a regular tetrahedron, Fig. A.1. Along these unit-vectors, a water molecule interacts

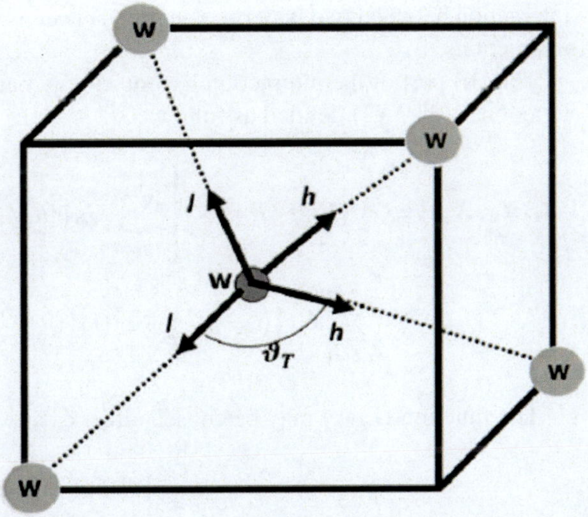

Fig. A.1 The geometry of a BN3D water model. Four unit-vectors pointing to the vertices of a regular tetrahedron. The angle between two vectors is the tetrahedral angle ϑ_T

A. Ben-Naim, *Solvent-Induced Interactions and Forces in Protein Folding*, Physical Chemistry in Action, https://doi.org/10.1007/978-3-031-31871-9

with other water molecules in a way similar to what we know as HBing. The angle between any two pairs of unit-vectors is the tetrahedral angle $\vartheta_T = 109.47°$. This angle is defined by:

$$\vartheta_T = \arc\cos\left(-\frac{1}{3}\right) = 109.47° \tag{A.1}$$

For details, see Appendix A of Ben-Naim [2]. Note that this angle is also the same as the angle between two oxygen atoms, the O–O bond, in ordinary ice structure.

The water-water pair-potential is now written as a sum of two terms:

$$U(X_1, X_2) = U_{LJ}(R_{12}) + \in_{HB} G(X_1, X_2) \tag{A.2}$$

The first term, $U_{LJ}(R_{12})$, is the Lennard–Jones pair potential which describes the interaction between the two neon atoms. The parameter ε_{HB} is the HB energy $\left(\in_{HB} \approx -6 \text{ kcal mol}^{-1}\right)$ and $G(X_1, X_2)$ is essentially a geometric function defined below, which has the following property:

$$G(X_1, X_2) = \begin{cases} 1, \; when\ the\ two\ water\ molecules\ form\ a\ \text{HB} \\ 0, \; otherwise \end{cases} \tag{A.3}$$

The condition: "*When the two water molecules form a* BB" will be discussed below.

In 1974, when this model was suggested, the water-water pair potential included also a dipole–dipole interaction to account for the long-range interaction between water molecules. This term might be important for a theory of liquid water where the long-range interaction might contribute significantly to the total potential energy of the system. However, in this book, we are interested only in the short-range interaction between two water molecules. This consists mainly of the LJ and the HB interactions.

The HB part of the interaction has the energy parameter ε_{HB}, and a *geometrical* function $G(X_1, X_2)$ defined as follows:

$$G(X_1, X_2) = G_{\sigma'}(R_{ij} - R_H) \times \left\{ \sum_{\alpha,\beta=1}^{2} G_\sigma\left[\left(h_{i\alpha} \cdot u_{ij}\right) - 1\right] G_\sigma\left[\left(l_{j\beta} \cdot u_{ij}\right) + 1\right] \right.$$
$$\left. + \sum_{\alpha,\beta=1}^{2} G_\sigma\left[\left(l_{i\alpha} \cdot u_{ij}\right) - 1\right] G_\sigma\left[\left(h_{j\beta} \cdot u_{ij}\right) + 1\right] \right\} \tag{A.4}$$

The function $G_\sigma(x)$ may be chosen either as a step-function of the form:

$$G_\sigma(x) = \begin{cases} 1 \; for \; |x| < \sigma \\ 0 \; for \; |x| \geq \sigma \end{cases} \tag{A.5}$$

or as an un-normalized Gaussian function defined by:

$$G_\sigma(x) = \exp\left(-x^2/2\sigma^2\right) \tag{A.6}$$

In (A.4), $h_{i\alpha}$ and $h_{i\beta}$ are two unit-vectors along which the water molecule can form HB as a donor (these two directions are the OH directions in ice, differing slightly from the OH direction in a single water molecule. See Ben-Naim [2]. $l_{i\alpha}$ and $l_{i\beta}$ are the unit-vectors along which a water molecule can form HB as an acceptor (these directions are similar to the direction of the lone-pair of electrons in water, Ben-Naim [1.2]).

$\boldsymbol{u}_{ij}$ is a unit-vector along the vector $\boldsymbol{R}_j - \boldsymbol{R}_j$ and is defined by:

$$\boldsymbol{u}_{ij} = \frac{\boldsymbol{R}_{ij}}{R_{ij}} \tag{A.7}$$

The three factors in Eq. (A.4) may be expressed by Gaussian functions. The first function gets its maximal value when the distance R_{ij} is about the HB distance in ice (or the O–O distance in ice) which is $R_H = 2.76$ Å. We assume that this function can be either continuous or discrete. Each of the factors in the curled brackets can also be either continuous or discrete. It attains its maximum value whenever one unit-vector, say, $h_{i\alpha}$ is in the direction of the unit-vector, $\boldsymbol{u}_{ij}$, and at the same time the unit-vector $l_{j\beta}$ points towards $-\boldsymbol{u}_{ij}$.

The whole expression in Eq. (A.4) looks complicated, but it is quite simple. Consider first the function $G_{\sigma'}\left(R_{ij} - R_H\right)$, where R_H is the water-water distance at which we expect a hydrogen bond to be formed. This is the distance we denoted by $R_H = 2.76$ Å. This function attains its maximal value of unity when the distance between the two molecules is R_H. It drops to zero [either abruptly as in (A.5) or continuously as in (A.6)] for $\left|R_{ij} - R_H\right| > \sigma'$. The next terms in Eq. (A.4) contain conditions on the relative orientation of the pair of molecules. The factor $G_\sigma\left[\left(h_{i\alpha} \cdot \boldsymbol{u}_{ij}\right) - 1\right]$ attains its maximum value (of unity) whenever the unit-vector $h_{i\alpha}$ (i.e. the donor arm $h_{i\alpha}(with\alpha = 1, 2)$ of the ith molecule) is in the direction of the unit-vector $\boldsymbol{u}_{ij} = \boldsymbol{R}_{ij}/R_{ij}$. Similarly, $G_\sigma\left[\left(l_{j\beta} \cdot \boldsymbol{u}_{ij}\right) + 1\right]$ attains its maximum value (of unity), whenever the direction of the acceptor arm $l_{j\beta}$ is in the direction $-\boldsymbol{u}_{ij}$. Thus, the product of these three Gaussians attains a value close to unity only when the three conditions are fulfilled simultaneously; R_{ij} is about R_H, the direction of $h_{i\alpha}$ is about that of $\boldsymbol{u}_{ij}$, and the direction of $l_{j\beta}$ is about that of $-\boldsymbol{u}_{ij}$. We call such a configuration a "favorable" one for the formation of a HB. When all of these three conditions are fulfilled, we say that a HB is formed, and the interaction energy of the pair of molecules is about in_{HB}. The eight terms in the curly brackets of (A.4), corresponding to the eight possible configurations which we considered to be favorable for HB formation. In practical applications, we consider the variances σ and σ' as adjustable parameters. In most applications we also choose the square-well function (A.5), which leads to an on–off definition of a HB. The continuous definition of the HB potential using (A.6) provides a gradual change from a HBed to a non-HBed pair of water molecules. This is clearly a more realistic view of the hydrogen bond part of the interaction.

In some applications, the function $G(X_i, X_j)$ is further simplified, by disregarding the difference between a donor and an acceptor arm, so that all the "arms" are considered to be equivalent. Thus, instead of eight terms in (A.4), corresponding to the eight pairings of a donor and acceptor, we take sixteen terms corresponding to the sixteen pairings of one arm on the ith molecule with one "arm" on the jth molecule. This simplified version of the potential was referred to as the primitive model for water Ben-Naim [2].

The Second Virial Coefficients of Water

The virial coefficients of any substance are defined as the coefficients in the density expansion of the pressure:

$$\frac{P}{kT} = \rho + B_2(T)\rho^2 + B_3(T)\rho^3 + \cdots \tag{A.8}$$

The second virial coefficient for non-spherical molecules is given by:

$$B_2(T) = -\frac{1}{16\pi^2}\int \{\exp[-\beta U(X_1, X_2)] - 1\} dX_2 \tag{A.9}$$

There were several estimates of the second virial coefficient of water. There are some also experimental data. For details, see Ben-Naim [2], Chap. 2. In this book, we shall use the experimental data to estimate the factor $\exp[-\beta \in_{HB}]\Delta X$ which will be used in Appendix B, to estimate the solvation Gibbs energy of water in water.

Here we use an empirical equation, for the second virial coefficient as a function of temperature, derived by Keyes (1958):

$$B_2(T) = \{2.062 - (2.9017 \times 10^3/T)\exp[1.7095 \times 10^5/T^2]\} \times 18\ \text{cm}^3\text{mol}^{-1} \tag{A.10}$$

For very high temperatures we have the limiting value:

$$B_2(T \to \infty) \approx 37.1\ \text{cm}^3/\text{mol} \tag{A.11}$$

We now rewrite the second virial coefficient in two parts:

$$B_2(T) = -\frac{1}{16\pi^2}\int \{\exp[-\beta U(X_1, X_2) - 1]\} dX_2$$

$$\approx -\frac{1}{16\pi^2}\left\{(-1)\int dX_2 + \left[\exp(-\beta \in_{HB}) - 1\right]\int \underset{HBregion}{\cdots} \int dX_2\right\} \tag{A.12}$$

The first term, on the right-hand side of Eq. (A.12), corresponds to the region of integration in which the geometric function is zero, i.e., $G(X_1, X_2) = 0$ and $U_{HS}(R_{12}) \to \infty$. In this region, the integrand is (– 1). If we choose the diameter of a water molecule to be $\sigma_W \approx 2.8$ Å, then we obtain:

$$\frac{1}{16\pi^2}\int dX_2 = \frac{1}{2}\int_0^{\sigma_W} 4\pi R^2 dR = \frac{4\pi\sigma_W^3}{6} \approx 23 \text{ cm}^3\text{mol}^{-1} \tag{A.13}$$

This is about half of the excluded volume between two hard-spheres of diameter 2.8 Å. The second part of the integral in Eq. (A.12) is approximated as follows:

We assume that the HB can be formed only from a small region ΔX, in the configurational space, where the integrand is nearly constant. There are eight such regions from which water molecule 2 can form a HB with water molecule 1. We define the HB region per arm, ΔX as:

$$8\Delta X = \int \underset{HBregion}{\cdots} \int dX_2 \tag{A.14}$$

Using (A.13) and (A.14), we can rewrite (A.12) as:

$$B_2(T) = \frac{4\pi\sigma_W^3}{6} - \frac{8}{2}\{\exp(-\beta \in_{HB}) - 1\}\frac{\Delta X}{8\pi^2} \tag{A.15}$$

The value of ΔX may be estimated from the experimental value of the second viral coefficient. For instance, at $T = 25$ °C, the value of B_2 is estimated from Eq. (A.10), as:

$$B_2(T = 298 \text{ K}) = -116\ 5 \text{ cm}^3\text{mol}^{-1} \tag{A.16}$$

Using the value of the first term in (A.12), we estimated in Eq. (A.13), we obtain:

$$\frac{\Delta X}{8\pi^2}\left[\exp(-\beta \in_{HB}) - 1\right] = \frac{28 + 1165}{4} \approx 298.25 \text{ cm}^3\text{mol}^{-1} \tag{A.17}$$

From this equation we can estimate a value of ΔX for any choice of $\in_{HB}$. Clearly, the stronger we choose the HB energy, the smaller the HB region necessary to fit the experimental value of the second viral coefficient. In Appendix B, we will use the value of the product of the two factors on the left-hand side of (A.17), to estimate the solvation Helmholtz (or Gibbs) energy of a water molecule in liquid water.

References

1. Ben-Naim A (1974) Water and aqueous solutions, introduction to a molecular theory. Plenum Press, New York
2. Ben-Naim A (2009) Molecular theory of water and aqueous solutions, part I: understanding water. World Scientific, Singapore

Appendix B
Solvation Thermodynamics of Water in Liquid Water

Traditionally, the concept of solvation and the corresponding thermodynamic quantities were studied for solute molecules in a very dilute solution in a solvent, Ben-Naim [1, 2]. However, with the new definition of solvation, first published in 1978, [6], this concept could be applied to any molecule in any solvent including the solvation of water in pure liquid water. The latter was referred to as self-solvation. This appendix is about the self-solvation of water, in liquid water. The solvation Gibbs energy of water in liquid water is an important quantity in the study of hydrophilic interaction.

For the present discussion, we treat a water molecule as a rigid molecule, the configuration of which is characterized by the six coordinates $\boldsymbol{X} = \boldsymbol{R}, \boldsymbol{\Omega}$. (three locations and three orientations).

We define the pseudo-chemical potential of water in liquid water as:

$$\mu_w^{*l} = -kT \ln \frac{\Delta^l(T, P, N; \boldsymbol{X}_0)}{\Delta^l(T, P, N)} \tag{B.1}$$

Here, Δ^l is the (T, P, N)-partition function (PF), and μ_w^{*l} is the change in Gibbs energy for the process of adding one water molecule at a fixed configuration $\boldsymbol{X}_0$ into a liquid with N water molecules at a constant temperature, T and pressure, P.

Similarly, one can define the pseudo-chemical potential of water in an ideal-gas phase as:

$$\mu_w^{*ig} = -kT \ln \frac{\Delta^{ig}(T, P, N; \boldsymbol{X}_0)}{\Delta^{ig}(T, P, N)} \tag{B.2}$$

With these two definitions in Eqs. (B.1) and (B.2), we proceed to define the solvation Gibbs energy of water in pure liquid water as:

$$\Delta G_w^* = \mu_w^{*l} - \mu_w^{ig} \tag{B.3}$$

A. Ben-Naim, *Solvent-Induced Interactions and Forces in Protein Folding*, Physical Chemistry in Action, https://doi.org/10.1007/978-3-031-31871-9

This definition was introduced in the late 1980s. It has several advantages over the "conventional" standard Gibbs energy of solvation. For details see Ben-Naim [1].

Since in an ideal gas phase there are no interactions between the molecules, the pseudo chemical potential, in an ideal gas phase can be written as:

$$\mu_w^{*ig} = -kT \ln q_w \tag{B.4}$$

where q_w is the *internal* PF of a single water molecule (including, vibrational, rotational and electronic, but not the momentum PF). See Ben-Naim [2, 3].

We assume that the internal partition function of a water molecule is nearly the same in the liquid and in the ideal-gas phase. Hence, the solvation Gibbs energy is due to the *binding* energy of one molecule to the rest of the system, see below. We write the statistical mechanical expression for the solvation Gibbs energy as:

$$\Delta G_w^* = -kT \ln\langle \exp[-\beta B_w(\boldsymbol{X}_o)]\rangle_o \tag{B.5}$$

In Eq. (B.5), $B_w(\boldsymbol{X}_o)$ is the binding energy of the water molecule at $\boldsymbol{X}_o$, defined by:

$$B_w(\boldsymbol{X}_o) = U(\boldsymbol{X}^N, \boldsymbol{X}_o) - U(\boldsymbol{X}^N) \tag{B.6}$$

For systems with pairwise additive interactions Eq. (B.6) reduces to:

$$B_w(\boldsymbol{X}_o) = \sum_{i=1}^{N} U(\boldsymbol{X}_i, \boldsymbol{X}_o) \tag{B.7}$$

Note that the binding energy is defined as the change in the total interaction energy of the system, being at a specific configuration $\boldsymbol{X}^N$ when we add a new water molecule at a fixed configuration $\boldsymbol{X}_o$, Fig. B.1a For systems with pairwise additive interactions, the binding energy is simply the sum of all the interaction energies of the added particle to the *N* particles being at some specific configuration, Fig. B.1b.

The average denoted $\langle\ \rangle_o$ is over all the configurations of the *N* water molecules with the probability distribution:

$$\Pr(\boldsymbol{X}^N) = \frac{\exp[-\beta U(\boldsymbol{X}^N)]}{\int \exp[-\beta U(\boldsymbol{X}^N)] d\boldsymbol{X}^N} \tag{B.8}$$

B.1 Experimental Values of ΔG_w^*

The solvation Gibbs energies of water in water may be obtained from data on the densities of water in the liquid and in the gaseous phase at equilibrium.

At equilibrium, the chemical potentials of water in the two phases must be equal. Hence, we write the equilibrium condition as:

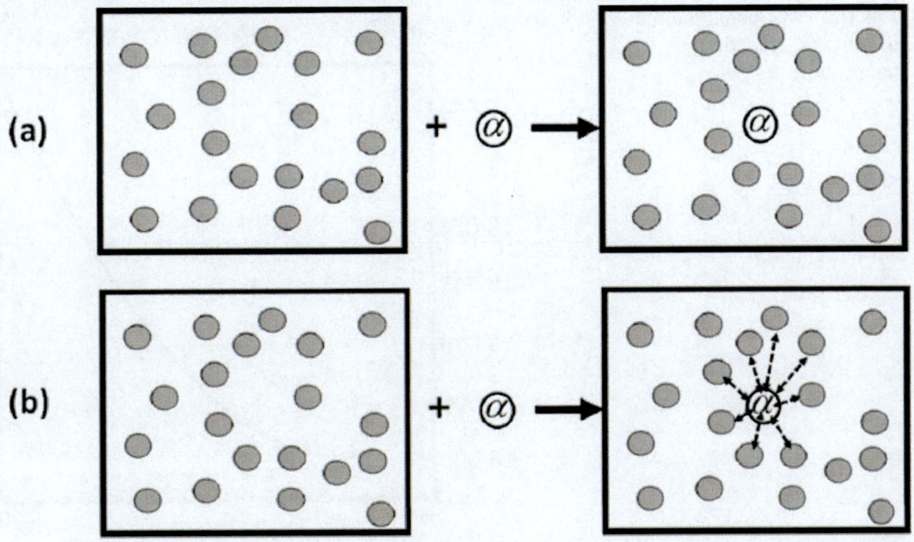

Fig. B.1 **a** Definition of Binding energy. The change in the total interaction energy in the system, cause by the addition of a solute. **b** The same as in (**a**) but with the assumption of pairwise additivity

$$\mu_w^l - \mu_w^g = \mu_w^{*l} - \mu_w^{*g} + kT\ln\left(\rho_w^l/\rho_w^g\right)_{eq} = 0 \tag{B.9}$$

From which we get the equality:

$$\mu_w^{*l} - \mu_w^{*g} = -kT\ln\left(\rho_w^l/\rho_w^g\right)_{eq} \tag{B.10}$$

On the right-hand side, we have the ratio of the densities in the two phases at equilibrium. On the left-hand side, we have approximately the Gibbs energy of solvation of water in water. To see this, we rewrite the left-hand side of Eq. (B.10) as:

$$\mu_w^{*l} - \mu_w^{*g} = \mu_w^{*l} - \mu_w^{*ig} + \mu_w^{*ig} - \mu_w^{*g} = \Delta G_w^{*l} - \Delta G_w^{*g} \tag{B.11}$$

where ΔG_w^{*g} is the Gibbs energy of solvation of water into the (real) gas (g) at equilibrium with liquid water. Thus, the quantity on the left-hand side of Eq. (B.10) is the difference between the solvation Gibbs energy in the liquid, and in the vapor above the liquid at equilibrium. In the range of temperatures between 0 °C and 100 °C we can safely assume that, the vapor is an ideal gas (*ig*), in this case, we can write the approximation:

$$\Delta G_w^{*g} \approx \Delta G_w^{*ig} = 0 \tag{B.12}$$

Hence, we have:

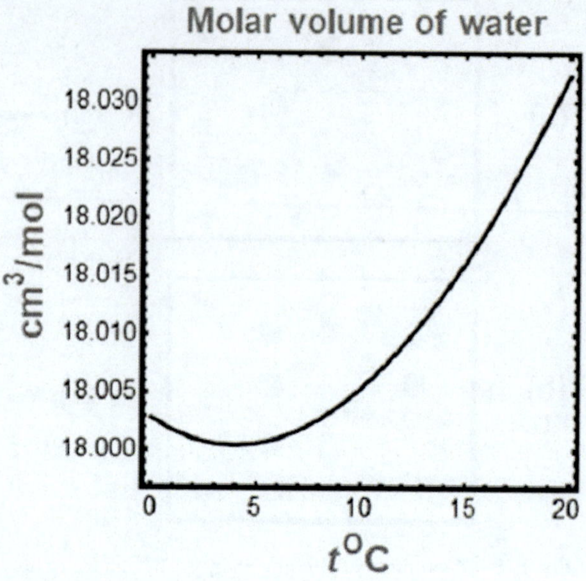

Fig. B.2 The molar volume of water as a function of temperature

$$\Delta G_w^{*l} = kT \ln\left(\rho_w^l / \rho_w^g\right)_{eq} \tag{B.13}$$

We can further estimate the density of the water in the gaseous phase as:

$$\rho_w^g \approx P_w / kT \tag{B.14}$$

where P_w is the vapor pressure of the water at the temperature T. Thus, we have the final approximate expression:

$$\Delta G_w^{*l} \approx kT \ln\left(P_w / kT \rho_w^l\right)_{eq} \tag{B.15}$$

where on the right-hand side we have the experimental quantities P_w, ρ_w^l and T at equilibrium. Figure B.2 shows the molar volume of water as a function of temperature between zero to twenty degrees centigrade, $v_{\mathrm{w}} = 1/\rho_w^l$.

Figure B.3 shows the values of the self-solvation Gibbs energy of water and heavy water in the range of temperatures between 0 °C and 100 °C. Note that the solvation Gibbs energies of D_2O are systematically lower (more negative) than those of H_2O. This is consistent with the known fact that the HB energy in D_2O is stronger than in H_2O

From these data one can obtain the entropy, the enthalpy shown in Figures B.4 and B.5. Other quantities of solvation of water in pure water are also obtainable from these data, see Ben-Naim [5], Chap. 1.

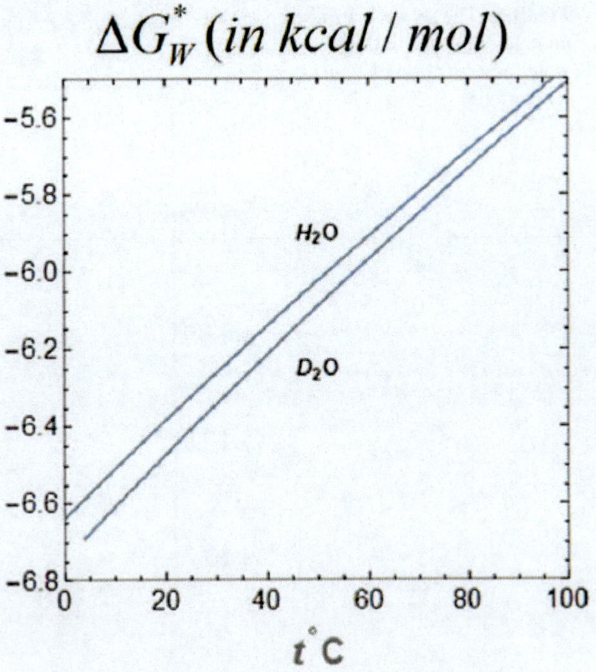

Fig. B.3 The Gibbs energy of solvation of water in liquid water as a function of temperature

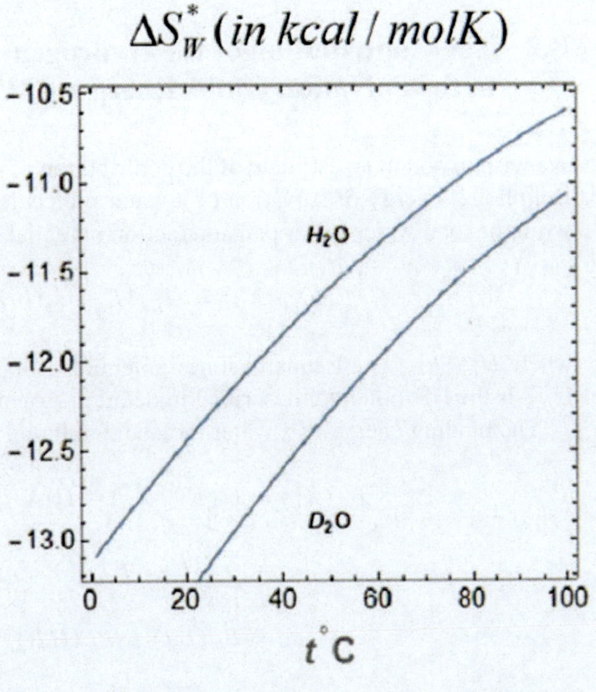

Fig. B.4 The self-solvation entropy of water in liquid water as a function of temperature

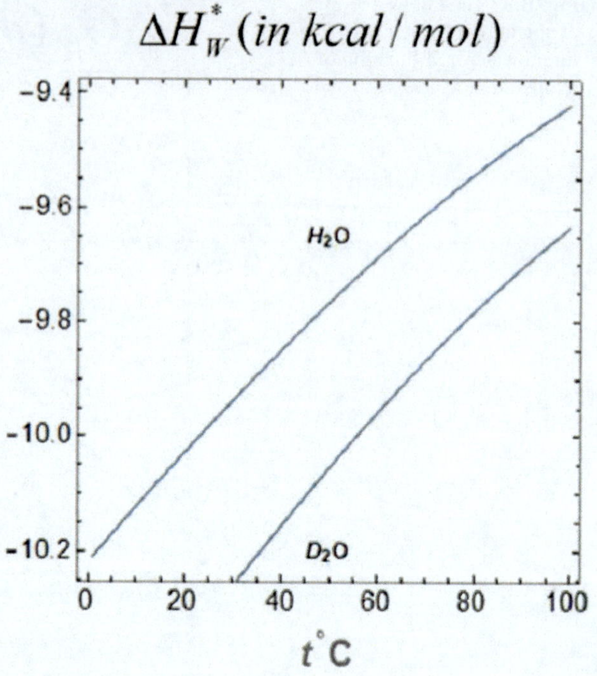

Fig. B.5 The self-solvation enthalpy of water in liquid water as a function of temperature

B.2 The Contribution of the Hydrogen-Bonding to the Solvation Gibbs Energy of Water in Water

We wish to obtain an estimate of the contribution of the HBing to the total Gibbs (or Helmholtz) energy of solvation of a water molecule in liquid water. To do that we write the total water-water pair interaction potential (see Appendix A) as:

$$U(\mathbf{X}_i, \mathbf{X}_j) = U_{LJ}(R_{ij}) + U_{HB}(\mathbf{X}_i, \mathbf{X}_j) \tag{B.16}$$

where $U_{LJ}(R_{ij})$ is a Lennard–Jones potential with parameters as for neon, and the U_{HB} is the HB-potential described in detail in Appendix A.

The binding energy of a water molecule in liquid water is written as:

$$\begin{aligned} B_w(\mathbf{X}_o) &= U(\mathbf{X}^N, \mathbf{X}_o) - U(\mathbf{X}^N) \\ &= \sum_{i=1}^{N} \{U_{LJ}(R_{o,j}) + U_{HB}(\mathbf{X}_o, \mathbf{X}_j)\} \\ &= B_w(LJ) + B_w(HB) \end{aligned} \tag{B.17}$$

With this binding energy, we can write the Gibbs energy of solvation as:

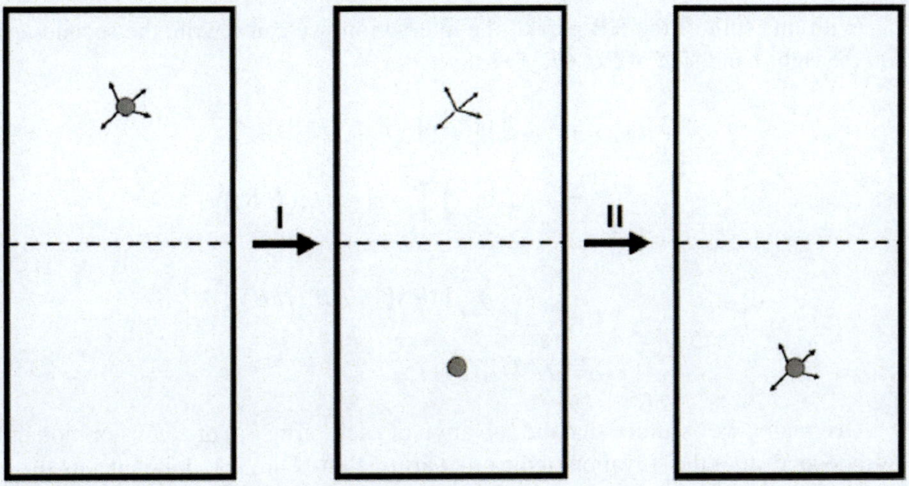

Fig. B.6 Two-step solvation process of a water molecule: Adding only the LJ part of the potential (I), and adding only the HB part of the potential (II)

$$
\begin{aligned}
\Delta G^*_w &= -kT \ln\langle \exp[-\beta B_w(X_o)]\rangle_o \\
&= -kT \ln\langle \exp[-\beta B_w(LJ) - \beta B_w(HB)]\rangle_o \\
&= -kT \ln\langle \exp -\beta B_w(LJ)\rangle_o - kT \ln\langle \exp -\beta B_w(HB)\rangle_{LJ} \\
&= \Delta G^*_{LJ} + \Delta G^*_{HB|LJ}
\end{aligned} \tag{B.18}
$$

Note that we dropped the superscript (l) from ΔG^*_w. What we have done in Eq. (B.18), is simply to perform the process of solvation in two steps: First, we place a LJ particle at some fixed position in the liquid, then we add the HB-interaction. In the first step the work is the same as the solvation Gibbs energy of neon (in water). In the second step we "turn-on" the HB part of the potential, given that the LJ part has already been turned on. Figure B.6.

Note that the first average on the right-hand-side of Eq. (B.18) is over all the configurations of the N water molecules, with the distribution (B.8), i.e., before adding the LJ particle. The second average denoted $\langle\ \rangle_{LJ}$ is also over all the configurations of the N water molecules, but with the *conditional probability* of the N water molecules, given that a LJ particle is placed at a fixed position in the liquid.

For the first part, we can use the experimental data on the Gibbs (or Helmholtz) energy of solvation of neon in water.

For the second part, we further split the HB part of the binding energy as:

$$B_w(HB) = \sum_{i=1}^{4} B_i(HB) \tag{B.19}$$

where $B_i(HB)$ is the binding energy along the ith arm of the water molecule. See Appendix A.

With this split of the HB part of the interaction, we can rewrite the second term on the right-hand side of Eq. (B.18) as:

$$\begin{aligned}\Delta G^*_{HB|LJ} &= -kT\ln\langle\exp[-\beta B_w(HB)]\rangle_{LJ}\\ &= -kT\ln\ \langle\prod_{i=1}^{4}\exp[-\beta B_i(HB)]\rangle\ _{LJ}\\ &= -kT\sum_{i=1}^{4}\ln\langle\exp[-\beta B_i(HB)]\rangle_{LJ}\\ &= 4\Delta G^*_{a|LJ}\end{aligned} \tag{B.20}$$

Originally, we assumed that the solvation of each "arm" (a) of the water molecule is independent of the solvation of the other arms (Ben-Naim [3]). Later, it was shown by simulation that this assumption is a good approximation (Predota et al. [4]). Thus, with this assumption the average over the product of the four functions in (B.20) is a product of the four averages. Therefore, the work involved in "turning on" the HB part of the interaction is simply equal to four times the work of "turning on" one arm. (Although there are two types of arms, donor and receptor, the solvation Helmholtz energy is assumed to be the same for each arm).

We can now use Eqs. (B.18) and (B.20) to estimate the contribution of one "arm" to the solvation Gibbs energy of a water molecule, which we write as:

$$\Delta G^*_{HB|LJ} = 4\Delta G^*_{a|LJ} = \Delta G^*_w - \Delta G^*_{\text{neon}} \tag{B.21}$$

On the right-hand side of Eq. (B.21), we have two experimental quantities: The solvation Gibbs energy of neon in water, and water in water (see Ben-Naim [1]). The values of these solvation Gibbs energies for 25 °C and at one atmospheric pressure, are:

$$\begin{aligned}\Delta G^*_{\text{neon}}(in\ H_2O) &= 2.67\ \text{kcal/mol}\\ \Delta G^*_{\text{neon}}(in\ D_2O) &= 2.63\ \text{kcal/mol}\\ \Delta G_{H_2O}(in\ H_2O) &= -6.32\ \text{kcal/mol}\\ \Delta G_{D_2O}(in\ D_2O) &= -6.41\ \text{kcal/mol}\end{aligned} \tag{B.22}$$

Note that the solvation Gibbs energy of neon in water is positive, whereas that of water in water is much larger and negative.

From the data in (B.22) we can get the conditional Gibbs energy of solvation of one "arm" (see Eq. 3.20)) as:

For H_2O:

$$\Delta G^*_{a|LJ} = \frac{\Delta G^*_{H_2O} - \Delta G^*_{\text{neon}}}{4} = -\frac{8.99}{4} = -2.25\ \text{kcal/mol} \tag{B.23}$$

For H_2O:

$$\Delta G^*_{a|LJ} = \frac{\Delta G^*_{D_2O} - \Delta G^*_{\text{neon}}}{4} = -\frac{9.04}{4} = -2.26 \text{ kcal/mol} \quad \text{(B.24)}$$

We shall use these values in Chap. 4 to estimate the $H\phi I$ interaction at some specific configuration of water molecules.

It should be noted that had we done the same calculation for H_2S and H_2Se we would get very much smaller values for the solvation Gibbs energy (see Ben-Naim [5], Chap. 2).

B.3 Theoretical Estimate of ΔG^*_w

In this section, we shall estimate ΔA^*_w instead of ΔG^*_w. We do that mainly for convenience. The two quantities differ by the quantity $P\Delta V^*_w$ which is negligibly small compared with the value of ΔG^*_w (or ΔA^*_w) (Ben-Naim [1]).

We start with the definition of the conditional solvation Helmholtz energy of one arm given that the Lennard Jones part of the pair potential has been "turned on."

$$\exp[-\beta \Delta A^*_{a|LJ}] = \int \Pr(X^N|LJ) \exp[-\beta B_a] dX^N \quad \text{(B.25)}$$

$\Pr(X^N|LJ)$ is the conditional probability density of finding the configuration $\boldsymbol{X}^N$ of all the N water molecules, given a LJ particle at some fixed position, say $\boldsymbol{R}_o$. B_a is the binding energy of one "arm" of a water molecule with all the N water molecules being at a specific configuration $\boldsymbol{X}^N$ (this is what we denoted $B_i(HB)$ in Eq. (B.19).

The integration in (B.25) is over all the configurations of the N water molecules. We can divide this range of integrations into two parts:

(1) All the configurations of the water molecules, for which one water molecule is HBed to the arm a.
(2) All other configurations of the water molecules, i.e. all configurations for which no water molecule is HBed to the arm a. Thus, we write the integral on the right-hand-side of (B.25) as:

$$\int \Pr(X^N|LJ) \exp[-\beta B_a] d\boldsymbol{X}^N = \int_{(1)} \cdots d\boldsymbol{X}^N + \int_{(2)} \cdots d\boldsymbol{X}^N \quad \text{(B.26)}$$

The ellipsis "$\cdots$" in the integrals on the right-had side of (B.26) means the same integrand as on the left-hand side. Since an "arm" of a water molecule can form at most one HB with a water molecule, and since the range of integration in (1) is over all configurations of all the water molecules for which there is *one* (any one) water molecule forming HB we can further write the first integral on the right-hand side of (B.26) as:

$$\int_{(1)} \Pr(X^N|LJ)\exp[-\beta B_a]dX^N$$
$$= \sum_{i=1}^{N} \int_{(i)} \Pr(X^N|LJ)\exp[-\beta B_a(a, X_i)]dX^N$$
$$= N \int_{(i=1)} \Pr(X^N|LJ)\exp[-\beta B_a(a, X_i)]dX^N \quad \text{(B.27)}$$

In the first equality on the right-hand-side of Eq. (B.27) we split the range of integration (1) into a sum of *N* integrals: The range of integration denoted (i) means that the integration is over all the configurations for which a specific water molecule *i* is HBed to the arm *a*. Clearly, to get the integral on the left-hand-side of (B.27) we have to sum over all the ranges of integrations: "Water numbered 1 is HBed to *a*," "water numbered 2 is HBed to *a*" … "water numbered *N* is HBed to *a*."

Since all the water molecules are equivalent we replace the sum over *i*, by *N* equal integrals, and take one representative term from the sum, say, the one with index ($i = 1$).

At this stage we have on the right-hand-side of (B.27) an integral over all the configuration for which water number 1 is HBed to "arm" *a*. But when arm *a* is HBed to water number 1, the total binding energy is simply the HB-interaction energy of "arm" *a* with the water molecule at $\boldsymbol{X}_1$, i.e.

$$B_a(a, \boldsymbol{X}_1) = U^{HB}(a, \boldsymbol{X}_1) \quad \text{(B.28)}$$

Next, we turn to the second integral on the right-hand side of (B.26). The range of integration is over all the configurations of water molecules for which *no water molecule* is HBed to "arm" *a*. For all these configurations the binding energy B_a is zero. Therefore, we can rewrite the second integral on the right-hand side of Eq. (B.26) as:

$$\int_{(2)} \Pr(X^N|LJ)dX^N \quad \text{(B.29)}$$

This is the probability of the event "no water molecule is HBed to arm *a*." Therefore, we can write this probability as one minus the probability of the complimentary event, i.e."

$$\begin{aligned}
&\Pr(no\ water\ molecule\ HBed\ to\ a)\\
&\quad = 1 - \Pr(at\ least\ one\ water\ molecule\ HBed\ to\ a)\\
&\quad = 1 - \sum_{i=1}^{N} \Pr(\ water\ numbered\ i\ is\ HBed\ to\ a)\\
&\quad = 1 - N\ \Pr(water\ numbered\ 1\ is\ HBed\ to\ a)\\
&\quad = 1 - \int_{(1)} \Pr(X^N|LJ)dX^N
\end{aligned} \tag{B.30}$$

We can now rewrite the Eq. (B.25) as:

$$\exp[-\beta A^*_{a|LJ}] = 1 + N \int_{(1)} \{\exp[-\beta U^{HB}(a, \boldsymbol{X}_1) - 1]\} \Pr(\boldsymbol{X}^N|LJ)d\boldsymbol{X}^N \tag{B.31}$$

Note that $U^{HB}(a, \boldsymbol{X}_1)$ depends only on the coordinates of the added water at $\boldsymbol{X}_0$ and the water molecule numbered 1. Therefore, the integration over all the configurations of the $N-1$ water molecules may be carried out to obtain the conditional density of water molecules at $\boldsymbol{X}_1$, given a LJ particle at $\boldsymbol{R}_0$, i.e.

$$\int \Pr(\boldsymbol{X}^N|LJ)d\boldsymbol{X}^{N-1} = \rho_w(\boldsymbol{X}_1|LJ) \tag{B.32}$$

Hence, Eq. (B.31) is now rewritten as:

$$\exp[-\beta A^*_{a|LJ}] = 1 + \int_{(1)} \{\exp -\beta U^{HB}(a, \boldsymbol{X}_1) - 1\}\rho_w(\boldsymbol{X}_1|LJ)d\boldsymbol{X}_1 \tag{B.33}$$

In Eq. (B.33), we have an integral over all configurations of one specific water molecule (numbered 1).

We now assume that the HB interaction is discrete, i.e. a water molecule can form a HB with an arm only from a certain range of configurations denoted $\Delta\boldsymbol{X}$. This means that when the integration in (B.33) is carried out, whenever $\boldsymbol{X}_1$ is within the range $\Delta\boldsymbol{X}$ at $\boldsymbol{X}_{HB}$ the value of $U^{HB}(a, \boldsymbol{X}_1)$ is ϵ_{HB}, and it is zero for all the other configurations. Hence, we rewrite the final expression for $\Delta A^*_{a|LJ}$ as:

$$\exp[-\beta\Delta A^*_{a|LJ}] = 1 + [\exp(-\beta \epsilon_{HB}) - 1]\frac{\rho w}{8\pi^2} g_{w,LJ}(\boldsymbol{X}_1, \boldsymbol{R}_0)2\Delta\boldsymbol{X} \tag{B.34}$$

In (B.34) ϵ_{HB} is the HB energy, $g_{w,LJ}(\boldsymbol{X}_1, \boldsymbol{R}_0)$ is the pair correlation function between a LJ particle at $\boldsymbol{R}_0$ and a water molecule at a configuration $\boldsymbol{X}_1$ from which it forms HB with a. Recall that an arm can either be a donor or an acceptor. Therefore, the water molecule forming a HB with a can have two configurations of size ΔX for which the integrand in non-zero.

$\rho_w = 5.55 \times 10^{-2}$mol/cm^3 is the bulk density of water at 25 °C. We choose tentatively $g_{w,LJ}$ to be about 2 based on simulation results on the water neon pair correlation function. This value may be adjusted to obtain agreement with the experimental value [see Eq. (B.36) below].

The quantity $\exp[-\beta \in_{HB} -1]\Delta X/8\pi^2$ has been estimated from the second virial coefficient of water, see Appendix A. Its value is 298.25 cm^3/mol at 25 °C. With these parameters we can calculate $\Delta A^*_{a|LJ}$ at 25 °C to be:

$$\begin{aligned}\Delta A^*_{a|LJ} &\approx -kT\ln\left(1 + 298.25 \times 2 \times 2 \times 5.55 \times 10^{-2}\right)\\ &= -1.987 \times 10^{-3}\ln(67.21) \approx -2.49\ \text{kcal/mol}\end{aligned} \tag{B.35}$$

This value is a little larger than the experimental value of −2.25 kcal/mol (see Eq. B.23)). We can choose the value of $g_{w,LJ}(X_1, R_0) \approx 1.32$ (instead of 2) to obtain agreement with the experimental value, i.e.

$$\Delta A^*_{a|LJ} \approx -2.25\ \text{kcal/mol} \tag{B.36}$$

We shall use this value as the conditional solvation Helmholtz (or Gibbs) energy for one arm of a water molecule, given that the LJ part of the interaction potential is already "turned-on."

References

1. Ben-Naim A (1987) Solvation thermodynamics. Plenum Press, New York
2. Ben-Naim A (2006) Molecular theory of solutions. Oxford University Press, Oxford, UK
3. Ben-Naim A (1992) Statistical thermodynamics for chemists and biochemists. Plenum Press, New York
4. Predota M, Ben-Naim A, Nezbeda I (2003) On the independence of the solvation of interaction sites of a water molecule. J Chem Phys 118:6446–6454
5. Ben-Naim A (2009) Molecular theory of water and aqueous solutions, part I: understanding water. World Scientific, Singapore
6. Ben-Naim A (1978) Standard thermodynamics of transfer: Uses and misuses. J Phys Chem 82:792–803

References and Suggested Reading

1. Anfinsen CB (1973) Principles that govern the folding of protein chains. Science 181:223–230
2. Apelblat A, Manzurola E, Balal NA (2006) J Chem Thermodynamics 38:565
3. Ben-Naim A (1971) Statistical mechanical study of hydrophobic interaction, part I: interaction between two identical non-polar solute particles. J. Chem. Phys. 54:1387–1404
4. Ben-Naim A (1971) Statistical mechanical study of hydrophobic interaction, part II: Interaction among a set of M identical spherical and non-polar solute particles. J Chem Phys 54:3696–3711
5. Ben-Naim A (1972) Statistical mechanical study of hydrophobic interaction, part III: generalization and further applications. J Chem Phys 57:5257–5265
6. Ben-Naim A (1972) Statistical mechanical study of hydrophobic interaction, part IV: the behavior of the function Y(R) at short distances. J Chem Phys 57:5266–5269
7. Ben-Naim A (1990) Solvent effects on protein association and protein folding. Biopolymers 29:567–596
8. Ben-Naim A (1990) Solvent-induced forces in protein folding. J Phys Chem 94:6893–6895
9. Ben-Naim A (1991) Strong forces between hydrophilic macromolecules; implications in biological systems. J Chem Phys 93:8196–8210
10. Ben-Naim A (2008) A farewell to entropy. Statistical mechanics based on information. World Scientific, Singapore
11. Ben-Naim A (2014) Discover probability; how to use it, how to avoid misusing it and how it affects every aspect of your life. World Scientific, Singapore
12. Dill K (1990) Dominant forces in protein folding. Biochemistry 29:7133–7155
13. Dill KA, Chan HS (1997) From levinthal to pathways to funnels. Nat Struct Biol 4:10–19
14. Kauzmann W (1993) Reminiscences from a life in protein physical chemistry. Protein Sci 2:671–691
15. Schellman JA (1987) The thermodynamic stability of proteins. Annu Rev Biophys Biophys Chem 16:115–137
16. Yu YB, Privalov PL, Hodges RS (2001) Biophys J 81:1632
17. Wolynes PG (2001) Landscapes, funnels, glasses and folding: from metaphor to software. Proc Amer Phil Soc 145:555–563

A. Ben-Naim, *Solvent-Induced Interactions and Forces in Protein Folding*, Physical Chemistry in Action, https://doi.org/10.1007/978-3-031-31871-9

Index

A. Ben-Naim, *Solvent-Induced Interactions and Forces in Protein Folding*, Physical Chemistry in Action, https://doi.org/10.1007/978-3-031-31871-9